APHASIA and
BRAIN ORGANIZATION

APPLIED PSYCHOLINGUISTICS
AND COMMUNICATION DISORDERS

APPLIED PSYCHOLINGUISTICS AND MENTAL HEALTH
Edited by R. W. Rieber

PSYCHOLOGY OF LANGUAGE AND THOUGHT
Essays on the Theory and History of Psycholinguistics
Edited by R. W. Rieber

COMMUNICATION DISORDERS
Edited by R. W. Rieber

APHASIA AND BRAIN ORGANIZATION
Ivar Reinvang

A Continuation Order Plan is available for this series. A continuation order will bring delivery of each new volume immediately upon publication. Volumes are billed only upon actual shipment. For further information please contact the publisher.

APHASIA and BRAIN ORGANIZATION

Ivar Reinvang

Sunnaas Hospital and
Institute of Psychology
University of Oslo
Oslo, Norway

PLENUM PRESS • NEW YORK AND LONDON

Library of Congress Cataloging in Publication Data

Reinvang, Ivar.
  Aphasia and brain organization

  (Applied psycholinguistics and communication disorders)
  Bibliography: p
  Includes index
    1. Aphasia. 2. Cognitive disorders 3.Brain—Wounds and injuries—Complications
  and sequelae 4 Neuropsychology. I. Title II Series
  RC424.7 R45   1985   616 85'52   85-9545
  ISBN 0-306-41975-0

©1985 Plenum Press, New York
A Division of Plenum Publishing Corporation
233 Spring Street, New York, N.Y 10013

Printed in the United States of America

PREFACE

This book presents the work on aphasia coming out of the Institute for Aphasia and Stroke in Norway during its 10 years of existence. Rather than reviewing previously presented work, it was my desire to give a unified analysis and discussion of our accumulated data. The empirical basis for the analysis is a fairly large group (249 patients) investigated with a standard, comprehensive set of procedures.

Tests of language functions must be developed anew for each language, but comparison of my findings with other recent comprehensive studies of aphasia is faciliated by close parallels in test methods (Chapter 2). The classification system used is currently the most accepted neurological system, but I have operationalized it for research purposes (Chapter 3).

The analyses presented are based on the view that aphasia is an aspect of a multidimensional disturbance of brain function. Findings of associated disturbances and variations in the aphasic condition over time have been dismissed by some as irrelevant to the study of aphasia as a language deficit. My view is that this rich and complex set of findings gives important clues to the organization of brain functions in humans. I present analyses of the relationship of aphasia to neuropsychological disorders in conceptual organization, memory, visuospatial abilities and apraxia (Chapters 4, 5, and 6), and I study the variations with time of the aphasic condition (Chapter 8).

No study of aphasia is complete without an analysis of its clinicoanatomical basis. Testing the assumptions of the classical model of aphasia, I can only partly confirm them. My analyses reveal that in many cases interactions between several lesion sites are important in determining deficits that are often thought to have a more circumscribed clinicoanatomical basis.

In taking such a broad view of aphasia, my theoretical framework has been influenced by concepts from general systems theory. The theoretical chapters of the book (Chapters 1 and 9) present and develop this type of approach sufficiently to account for the main aspects of my findings and to suggest some new lines of investigation for the future.

I should like to acknowledge the help and support of several friends and colleagues. First of all, K. Sundet performed the statistical analyses and discussed all the statistical problems with me. P. Borenstein examined my CT scans and scored them in a standardized system. K. Willmes made available to me the system for analysis of CT scans used in the Aachen aphasia laboratory. He also advised me on problems of choosing appropriate statistical models. My wife, T. Bjorg, did the artwork for the book. A special word of thanks to M. Taylor Sarno, who read an earlier version of this monograph and gave me every help and encouragement to develop it for publication.

Finally, I must thank the Norwegian National Health Association for their support of my work during several years, including the time period during which the book was written.

IVAR REINVANG

CONTENTS

Chapter 3. *Types of Aphasia*

Chapter 4. *Selective Aphasias*

Chapter 5. *Memory and Learning Deficits*

Chapter 6. *Defects of Visual Nonverbal Abilities*

Chapter 7. *Localization of Lesion in Aphasia*

Chapter 8. *Recovery and Prognosis*

Chapter 9. *The Organized Response of the Brain to Injury*

APPROACHES TO THE STUDY OF APHASIA

1.1. Clinical and Theoretical Approaches

The study of aphasia may be motivated by clinical as well as theoretical considerations. It has been estimated that about 1 million people suffer from aphasia in the United States (Sarno, 1980). In Sweden, the incidence of aphasia has been estimated at 60 per 100,000 inhabitants per year (Broman, Lindholm, & Melin, 1967), and in Norway, Petlund (1970) estimated the prevalence at .09%. The most frequent cause of aphasia is stroke, which is itself a common disease in an elderly population. Whereas the risk of stroke in the fifth decade of life is .2%, the corresponding risk in the seventh decade is 2.0% (Marquardsen, 1969). Add the fact that 20% to 25% of stroke patients are initially aphasic (Brust, Schafer, Richter, & Bruun, 1976), and the magnitude of the clinical problems becomes striking. In this context, the need for practical and reliable methods of testing is apparent. A classification system with knowledge of associated neurological and neuropsychological deficits, prognosis, and underlying pathology is a prerequisite for sound treatment.

From a theoretical point of view, aphasia has, since the time of the founding papers of Broca (1861) and Wernicke (1874), presented a unique opportunity to study the relationship of the brain to higher mental functions. The theoretical problem may, however, be formulated in different ways, and different methodological approaches may be chosen.

A clinically based research strategy includes the following steps:

1. Observing naturally occurring associations and dissociations of symptoms. These groupings are referred to as syndromes, with the understanding that they are clinically useful "fictions."

2. Distinguishing between theoretically meaningful associations (phenomena associated because they reflect the same function) and theoretically meaningless associations (associations produced by theoretically uninteresting combinations of functions). The methods used include psychometric analysis, *post hoc* control of the lesion variable (as the size of the lesion is believed to be the most significant factor in producing spurious associations of symptoms), and experimental control of the task variable. The result is a structural analysis of the function involved.

An alternative research strategy is based on assumptions about the nature of language in the normal case.

It is fair to say that, recently, theoretically oriented efforts have had the goal of analyzing (decomposing) the cognitive-linguistic process into constituent subfunctions and assigning neural correlates to these subfunctions. In order to attain the goal of accounting for processing, the internal computational steps of subfunctions and their ordering must be specified for a given type of task.

Progress in linguistics has led to models of the subcomponents of the language function, and to possible rules for relating linguistic symbols to each other. The structural school of linguistics has influenced aphasiology through the works of Jakobson (1971), whereas in more recent times the transformational generative model of linguistics presented by Chomsky (1965) has been influential.

The term *neurolinguistics* (Hécaen & Dubois, 1971; Whitaker, 1971) stands for an interdisciplinary study of aphasia based on neurology and linguistics.

In an influential paper, Arbib and Caplan (1979) argued that neurolinguistics must make an effort to give a computational account of processing and that this can be approached by converging efforts of neurolinguistics, psycholinguistics, artificial intelligence, and neurophysiology.

In the summary of Caplan (1982), the first steps are

1. The level at which the nature of computation is expressed. With respect to human language, Marshall suggests that generative transformational theories of grammar provide a characterization of the structures relevant to language, that is, a characterization of the features of the mental object attained.

2. The level at which algorithms that implement a computation are characterized. Marshall suggests that work on parsing strategies, both implemented and based on the results of psychological experimentation, provides an example of the beginnings of a characterization of the psychological steps which are operative in the attainment of the linguistic structures of Level 1.

3. The level at which an algorithm is committed to particular mechanisms, which has been "the traditional preserve within psycholinguistics of the aphasiologist." (p. 423)

Clinical and theoretical motivations have been closely wedded in the history of the study of aphasia. It has been assumed that theoretical inferences could be drawn with confidence on the basis of clinical observation of the association and dissociation of phenomena, and that the syndrome is a significant unit for theoretical analysis. Only recently has the closeness or fruitfulness of this alliance been questioned. According to Marshall (1982),

there will be some models of brain organization within which the demands of clinical diagnosis and theoretical understanding pull in diametrically opposed directions. (p. 404)

For the neurolinguist, the unit of analysis is language, and assumptions about a language function are independently motivated from studies in linguistics and psycholinguistics. It does not follow, however, that aphasia *in toto* or subsets of aphasia phenomena are wholly interpretable as a failure in subfunctions or processing stages of the language function. It is interesting to follow the increasing divisions of clinical and neurolinguistic studies. In the 1970s there was optimism that the major clinical syndromes of aphasia could be given a neurolinguistic analysis referring to breakdown in major blocks of linguistic subfunctions (Caramazza & Berndt, 1978). The more recent attitude is that only selected aphasic symptoms, including agrammatism and some forms of dyslexia and agraphia, can be usefully studied, and then preferably in selected cases with "pure" defects.

Taking the stand that there are only two approaches to aphasia— the clinical, which takes the patient as an unanalyzed whole as its unit of study, and the theoretical, which takes the language function and nonclinical models of it as its units of study—grossly disregards a third approach.

This approach may be termed *neuropsychological* and takes the syndrome as its unit of analysis. It follows the step of the clinical research procedure as outlined above. When the outlines and divisions of a functional domain have been established with gross neurological correlates, then one of two options may be chosen. One is to say that this is as far as one can get in clinical group studies, and selected patients with more specific deficits offer the only opportunity to advance the study by clinical material or methods. Another option, however, is to say that syndromes are meaningful indicators of the multidimensional response of the brain to localized injury. They are indications of the organizing principles at work in the efforts of the whole brain to maintain optimal functioning, as much as they are indications of the contributions of the missing parts.

Although neurolinguistic analysis is strong on detailed functional analysis and specification of processing, it has difficulty justifying its selection of study material and its notion of *relevant case*. We have no theoretical metric for measuring the "pureness" of functional deficits, and it may well be that sharply delineated behavioral symptoms are the consequence of highly complex functional interactions. The findings from analyses of pure deficits can be applied to more complex cases only if one assumes that pure deficits can be concatenated without giving rise to strong interaction effects. (See Shallice, 1979, for a discussion of the problems and advantages of single-case studies.)

Neuropsychological analysis is weak on detailed function—and process analysis. The existence of some alleged syndromes may be questioned on empirical grounds (e.g., the criticism of Benton, 1961, of the Gerstman syndrome). Neuropsychological analysis has the advantage of not requiring an independently motivated theoretical model and provides much of the necessary framework for more sophisticated analysis by giving a normative background for gradation of performances. If the phenomena under study are interactive, then syndromes may reflect the emergent properties of factors combined in a larger system, and a description of the relationship of the variables for different parametric values is a necessary part of a complete theoretical analysis.

1.2. Historical Approaches

The present conceptions of aphasia date back to the continental European neurological tradition before and around the turn of the century. This tradition has a main stream, represented by Wernicke (1874) and Lichtheim (1885) and several tributaries with supplementary approaches (Marie, 1906; Jackson, as summarized by Head, 1915, 1926; Goldstein, 1948; Luria, 1970; Jakobson, 1971; Hécaen & Dubois, 1971). To a surprising degree, the mainstream of thinking around the turn of the century is still a dominant mode of thought (Benson & Geschwind, 1977). Because all classical thories are centered on concepts of localization of function, it is useful to give a more general characteristic of localization theory before discussing the controversies.

1.2.1. Localization Theory of Language-to-Brain Relation

No author can be taken as the foremost representative of localization theory. The following is the present author's summary of the essential features of the theory implied by authors who use terms like "language area" or "speech center" to describe the neurological basis of the language functions:

1. The brain contains areas with specialized functions, beyond the sensory and motor areas. Normally, one cerebral hemisphere contains all the structures necessary and sufficient for language. This hemisphere is said to be dominant (for language). Normally, the left hemisphere is dominant, but in some instances, these structures may be distributed between the hemispheres or may be located entirely in the right hemisphere.

2. Within the dominant hemisphere, there is also specialization, so that some areas are of critical importance to the language function and some are not. The structures necessary for language (language areas) are commonly believed to be cortical, and to be located in the temporal and frontal lobes. There are, however, different theories and formulations of which specific areas are important and how far their functions are differentiated.

3. Different parts of the language areas are specialized for different functions. Differently localized lesions in the language areas give rise to varied clinical syndromes. By focusing on the features

that show the most consistent relationships to the locus of injury, a definition of types of aphasia can be given. It is not assumed that all pathological performances in aphasia show a lawful relationship to the locus of injury. Alternative classifications built on principles other than clinicopathological correlation may be chosen but would have to prove their advantage for special purposes. Again, there are different alternative formulations of which are the major and minor aphasic syndromes and what is their specific relationship to the locus of injury.

4. Language areas have fiber connections with one another and with other areas. In the classical localization theories, these connections are believed to have very simple functions of transmitting stimuli, thereby triggering the activity characteristic of the area receiving the stimulus. More complex information on the results of previous stages of analysis may also be transmitted, thereby "adding" or integrating the activity of several connected areas before a motor response is emitted. This simple conception of the functioning of connecting fibers has led to their being named *association fibers* and to their areas of convergence being called *association areas*. Although it has not been done in classical localization theory, it is entirely possible, without abandoning localization theory altogether, to explore the hypothesis that association fibers have more complex functions than believed.

The localization theory has been criticized both on general conceptual grounds and with respect to some of its more specific statements about the nature of aphasia and the types of aphasic disturbances.

It is worthwhile to pause and note that none of these criticisms question the existence of a correlation between the type of aphasia and the locus of the lesion. Even authors often identified as antilocalizationalists, like Jackson (see Head, 1915), Marie (1906), Head (1926), and Goldstein (1948), never denied the existence of clinicopathological correlation.

1.2.2. Criticism of Basic Premises

The criticisms most often advanced may be summarized under three points:

1. It is impossible or unacceptable to try to localize normal language, a criticism stated forcefully by Jackson (see Head, 1915).
2. The mixture of behavioral and neurological terms of classifications is ill-conceived and confusing. This criticism, too, is closely connected with the work of Jackson (see Head, 1915).
3. The general form of the theory (connectionism or associationism) is outdated and has been shown to be inadequate. Both Head (1926) and Pribram (1971) have stated this argument forcefully.

Regarding the alleged nonlocalizability of normal language, it is appropriate to stress the difficulty of using observational clinical data as a basis for inference about normal processes. In particular it is unwise to name "centers" on the direct basis of lesion locus and symptom description. This is no more than to say that phrenology is outdated as a model of neuropsychological research. On the basis of observation that patients with certain lesions have difficulty in naming objects, we would be unwise in inferring that the locus of the lesion is normally the locus of object names. But assume that characteristics of this naming difficulty can be teased out further by experimental variation of conditions and can be shown to deviate from normal performance by certain parameters. We would have then a basis for hypothesizing an underlying process, which can then, again hypothetically, be related to a given neurological structure. The hypotheses may very well have implications that could also be tested on normal individuals by means of behavioral measures.

The criticism is correct if it is reformulated to say that no hypothesis assigning normal processes to given neurological structures should be accepted on clinical evidence alone. The declaration that language cannot be localized, however, seems to be an arbitrary conceptual decision that any function with a definite relation to a neural locus cannot be called *language*.

Mixing behavioral and neurological classification was called "psychoneurology" by Jackson (see Head, 1915). There is danger of tautological reasoning if concepts from one category are used to define those from another. If *frontal aphasia* is defined as the type of aphasia resulting from frontal injury, then the question of the frontal localization of this syndrome has already been settled by definition.

However, if care is taken to define behavioral categories in behavior terms and neurological categories in anatomical terms, then there should be no objection to studying the relationships between the two.

With the advent of more sophisticated statistical techniques, the question of the behavioral validity of aphasia types can be raised: Are there naturally occurring clusters of aphasia symptoms, and if so, do they correspond to the classically described types of aphasia? It must be recognized that the answer to this question depends on the patient group studied. The agent of injury may be such as to produce diffuse lesions, as in metabolic or anoxic lesions. It may also produce discrete lesions, but of several anatomically distinct structures not known to have a common function. This combination of lesions may well occur in cerebrovascular disease, where structures may be damaged together by virtue of having a common blood supply. In penetrating head injuries, again, the lesion may be discrete and well defined but may not follow the demarcation lines drawn by anatomy. Rather than injuring one well-defined anatomical structure completely, it may incompletely injure three.

Modern statistical studies started with Weisenburg and McBride (1935) and continued with Jones and Wepman (1961) and Schuell, Jenkins, and Carroll (1962). All these studies rejected classical classification schemes but are open to the criticism of lack of control of localization of the lesion. Recent studies by Goodglass and Kaplan (1972) and Kertesz and Phipps (1977) indicate that an extension and refinement of classificatory schemes within the framework of a classical clinicopathological model are a likely development.

I agree wholeheartedly with the critics of classical localization theory that associationism is an inadequate theory for explaining the complex activity of the nervous system. Hughlings Jackson was aware of this point. From his studies of epilepsy, he described a certain class of symptoms as "release" symptoms, that is, symptoms caused by a loss of inhibition. In his hierarchial model, the alleged loss of the propositional level of function and the emergence of automatic speech are the primary example of this type of deficit in aphasia. In modern times, several authors, among them Pribram (1971), have rejected associationism and have proposed more complex theoretical models.

As noted above (p. 6), localization theory does not presuppose associationism, although the two theories have been closely linked historically. The inadequacy of an associationist model of the brain, together with the criticisms discussed above, should not lead to abandoning the concept of localization of function. Localization theory should, however, be modified and modernized. The specific content of an adequate theory is largely unknown. It is, however, of some interest to discuss what general features an adequate theory must have.

1.2.3. Criticisms of Assumptions Regarding the Nature of Aphasia or the Types of Aphasic Disturbances

The following criticisms will be discussed:

1. Language cannot be distinguished from intelligence. Aphasia is symptomatic of a more general intellectual disturbance (Marie, 1906; Bay, 1962).
2. Different forms of aphasia do not exist; only aphasia with different additional disturbances exists (Bay, 1962; Schuell, Jenkins & Jimenez-Pabon, 1965).

Jackson (see Head, 1915) proposed that language is integrated in several levels of mental functioning. Aphasia is not a disturbance of an anatomically localized language mechanism or process; rather, it reflects mental regression from a "propositional" level of functioning to lower levels. Speech in emotional context is preserved, but propositional speech is lost. Other influential thinkers supported this position. It was adopted by Head (1926), who echoed Jackson's statement that an aphasic is in a certain sense "lame in his thinking," and by P. Marie (1906), who said that aphasia is "a special sort of intelligence defect." In modern times Bay (1962) has been a strong advocate of the view that a conceptual disturbance is inherent in aphasia.

The position of Wernicke (1874) on this issue was clear:

> The spoken and written name of an object is not a new attribute of the object. It is thus clearly different from the actual sensory memory images of the object. Only the latter make up the concept of the object. Disturbance of the concepts of things with which we deal in the process of thinking are always disturbances of intelligence. Disturbances of speech, on the contrary, cause difficulties only in the use of the conventional means of representation of the concepts. (p.63)

The consequence of this issue for research seems to be to question whether a consistent defect in "intelligence" can be found in aphasics. If so, it is necessary to postulate an inherent link between thought and language beyond the plausible assumption that the language disturbance makes an instrument for thought less available. A way of demonstrating a defect of intelligence is to show that, given a defect in performing a task with language material, it is possible to demonstrate the defect even if the verbal elements of the task are removed.

The available research on hemispheric asymmetry only partly supports the notion of material-specific functions of the two hemispheres (Milner, 1974; Gazzaniga & Ledoux 1978), and differences in the cognitive mode of operation of the two hemispheres must be considered. (For review, see Bradshaw and Nettleton, 1981.) Research on intelligence in aphasia, summarized in Lebrun and Hoops (1974), indicates some reduction in specific nonverbal tasks, but the role of the size of the injury in explaining such defects is uncertain. The evidence on the issue is not strong enough to lead us to abandon the theory of localized language function. The facts and their interpretation are discussed further in Chapter 6.

The second challenge to localization theory and the clinicopathological model is the question of whether different types of aphasia exist. The position taken by antilocalizationists is that different syndromes exist after differently localized lesions, but they should not be called different forms of aphasia. They should rather be seen as aphasia with different, added disturbances. Marie (1906) stated that Broca aphasia is the combination of aphasia and anarthria. This is the holistic interpretation of aphasia, which has also been popular in modern times through the work of Schuell et al. (1965). There is no doubt that, in aphasia, variations in performance can often be observed, so that some patients have disproportionate difficulties with speaking, writing, reading, or auditory analysis. Sometimes such variations determine the classification of the type of aphasia. Whether or not such disturbances of performance should be called disturbance of language is partly a conceptual question. Benson and Geschwind (1977) defined language as "perception of verbal sensory stimuli, integration of these stimuli with prior knowledge, and activation of verbal response-mechanisms" (p. 2). This definition obviously allows variation in performance with different sensory modalities or response modes to be classified as language disturbances. Opponents would presumably restrict their definition of language to a cognitive mechanism and would exclude perceptual and response factors. Data from aphasiology can contribute to a resolution of this conceptual question. If it can be shown that specific neural circuits exist for programming speech or analyzing language, then it would seem natural to let the definition of language, at least from the physiologist's point of view, include the function of those circuits. If, on the other hand, auditory language perception cannot be distinguished from auditory perception in general, and programming of speech cannot be distinguished from programming of other complex motor behaviors, then there seems little reason to include perceptual and response mechanisms in the definitions of language. Rather, they would have to be viewed as tools for implementing language. This discussion is taken up in Chapter 4.

The evidence from psycholinguistics, from the work on speech perception and dichotic listening (Studdert-Kennedy & Shankweiler, 1970), and from language pathology in connection with dyslexia (Marshall & Newcombe, 1973) seems to favor a view of specialized perceptual mechanisms for language. The holistic or purely cognitive conception of language does not find support in these studies.

1.3. Systems Theory Approach

General systems theory (von Bertallanfy, 1948) views organisms as organized wholes. It is antireductionistic in the sense that it agrees with the traditional Gestalt slogan "The whole is more than the sum of the parts." On the other hand, it is not holistic in a traditional sense. It regards analysis of wholes into consistent parts and their mutual relationship as a fruitful undertaking. It does aim to transcend and integrate the traditional positions of reductionism versus holism. General systems theory has found applications in several sciences, among others in brain science.

Conditions for considering a systems theory type of analysis are present when there is evidence of organized complexity, that is, a systematic relationship between many factors without simple one-to-one correspondences. It is especially suited to a situation in which different phenomena mutually interact without being ordered in a causal chain. Weiss (1969) suggested statistical criteria for the presence of systems effect, one criterion being that, in a system, the variance of the whole is less than the sum of variances of the components, indicating that coupling of components constrains the output of the ensemble. Denenberg (1978) advocated the use of systems theory when investigations based on ANOVA designs show prominent inter-action effects and an absence of main effects.

Systems theory may be applied to the neural level of analysis by considering the possibility that developing or maintaining the func-tion of a piece of neural tissue is dependent on interactions with other units of neural tissue.

Systems theory may also be applied to the behavioral level when the influences of several factors are highly interactive, so that a decom-position and parceling out of causative factors becomes less mean-ingful than describing the different possible functional states of the system and their necessary conditions.

Von Bertallanfy (1948) introduced some general concepts that will be of use in a further discussion of brain function. The brain shares with other living systems the property of openness, that is, the property of relating to an environment. The brain (and other open systems) tends to continually reorganize the relations of its compo-nents to achieve more optimal functioning, and this process may be under predominantly *primary* regulation (determined by the structure of the brain itself) or under *secondary* regulation (determined by feed-back from the environment). Development may go in the direction of less interaction and greater degree of independent functioning of components, called *segregation*. This is accompanied by a process of *mechanization* in which the functional mode of a component becomes fixed.

An advantage of systems theory is its suitability to conception-alizing and organizing complex data. It thus has considerable heuristic value in structuring a complex field of study. Its disadvantages are its emphasis on description rather than explanation and the ensuing difficulty of deriving testable predictions. General systems theory is not a testable theory but only a framework for developing more detailed theories within a specific field.

1.3.1 Applications of Systems Theory Concepts in Neuropsychology

A comprehensive review of this topic is beyond the scope of this chapter. Few authors refer explicitly to systems theory itself, and very many use concepts that are related to systems theory in one way or another. Here I focus especially on statements emphasizing the dynamic, interactive nature of brain function and the nature of symptoms as an organized response of the whole brain after injury.

One context in which the dynamic nature of function-to-localization relationships has been much debated is the study of the ontogeny of the cerebral lateralization of the language function. Lenneberg (1967) summarized the evidence and concluded that language is gradually lateralized after being initially bilaterally represented. Since then, it has been recognized that this view is overstated. There is evidence of early specialization of the left hemisphere for language (Dennis & Whitaker, 1977), and for several aspects of language, there are no indications of continuing lateralization. As pointed out by Moscowitch (1977), current tests may tap only very low levels of linguistic processing, and "whether higher order linguistic processes do indeed become progressively more lateralized with age is open to debate" (p. 204).

Selnes (1974) gave an extensive review of the role of the corpus callosum in establishing hemispheric specialization. With poorly developed cortico-cortical connections, reciprocal specialization develops only so far as structural asymmetries allow. With callosal agenesis, there is an indication of greater likelihood of bilateral language representation and a higher incidence of retardation in language development. The development of hemispheric specialization for language in normals may be related to relatively late myelinization of callosal fibers (Yakovlev & Lecours, 1967). Selnes (1974) commented:

> The evidence in support of the view that CC (corpus callosum) may be responsible for the establishment of language lateralization is thus not strong, but at least it has the advantage of being relatively easy to confirm or disconfirm. This view differs from the "traditional" inhibitory theory

in that the latter ascribes to the CC the role of a more or less permanent
mediator of inhibitory influences from the dominant hemisphere, while
the present theory views the CC as instrumental only for the establish-
ment of language lateralization. Once this has been accomplished, there
should no longer be any need for inhibitory influences. This view does
not exclude, of course, that the CC is functional in transfer of information,
in particular visual information, and learning between the two hemispheres, and also in securing mental unity. (p. 132)

This distinction drawn by Selnes between establishing and maintaining language lateralization is important in explaining differences
between the effect of lesions on mature and immature nervous systems.

In the mature nervous system, patients with callosal section
preserve left-hemisphere lateralization of speech control. It has been
noted, however, that the right hemisphere in these patients has good
comprehension of auditory verbal stimuli. This finding is in apparent
contrast to the global aphasia with poor comprehension resulting from
massive left-hemisphere injury in stroke patients. It may therefore be
that, even in the mature nervous system, some change in preserved
tissue (reduced differentiation) takes place with loss of callosal input.

Denenberg (1981) reviewed the evidence for hemispheric specialization and differentiation in animals. Numerous examples of differences in the effects of left- and right-sided injuries can be cited. Of
even greater significance in the present context is the evidence that
the two hemispheres are systemically coupled, that is, that the function of an intact brain is not simply the sum of activities in the two
isolated hemispheres. Some of the actions of one hemispherie on the
other are inhibitory. Add evidence that hemispheric specialization is
sensitive to early experience, and the need for assuming a dynamic
component both in establishing and maintaining hemispheric specialization is strongly supported:

It is hypothesized that homologous brain areas and their connecting
callosal fibers must be intact at birth, and must be intact throughout
development for lateralization to reach its maximum level. If there is either
hemispheric damage or callosal damage the brain will be less specialized
with respect to hemispheric differences. The hypothesis specifies two
homologous brain areas and their connecting fibers as the "unit" for the
development of lateralization. This is based on the assumption that such
a unit will act to maximize neural heterogeneity (i.e., lateralization) because
of hemispheric competition. (Denenberg, 1981, p. 18)

Brown (1979) was concerned with the mechanisms of symptom
formation and of recovery and concluded that

in pathology levels in language production appear as symptoms. A symptom reveals a stage in language production that is traversed in the realization of the normal utterance. Brain damage has the effect of allowing symptoms—contents from more preliminary levels—to come to the fore. There may also be a regression to a more preliminary level. Accordingly, a brain lesion does not disrupt a mechanism or a center where that mechanism is situated. Rather, it involves that structural level through which the (pathological) content is normally elaborated. (p. 141)

The idea of a small set of levels that organize the basic phenomena of perception, action, reaction, and language is clearly related to earlier ideas of a hierarchical organization of the brain (Jackson, 1878, on propositional vs. automatic language). In the model of Brown, levels are also hierarchically organized, and the output of one is the input into the next:

There is a resubmission of emerging abstract content at each hierarchical level to the same reiterated process—in other words, one process at multiple levels, rather than multiple processes at the same level. (Brown, 1979, p. 142)

This is called a *micro-genetic process*.

Kinsbourne advanced the concept of functional cerebral space (Kinsbourne & Hicks, 1978), which makes the assumption that the brain is a highly linked neuronal network:

The programming of a particular continuous activity involves not only the cerebral locus at which the programming is accomplished, but also involves, by spread of activation, a large proportion of the total cerebral space, the amount occupied being greater the more closely the operator's performance approximates the maximum of which he is capable. (p. 346)

These authors explored the dimensions of functional cerebral space by studying the performance of concurrent tasks:

According to this model, if a single cerebral programme is being developed, the completed programs are facilitated to a greater extent at functionally closer loci than at functionally distant loci and, successively, transfer of training is greater to the closer locus. In contradiction, unrelated motor programs can be run concurrently most effectively if based on neural activity in loci functionally remote from each other. (p. 347)

The concept has been used in studies of developing lateralization of language by examining the interference between speech and left- or right-handed activities. This research has concluded that lateralization of spoken language in right-handers is established before 3 years of age (White & Kinsbourne, 1980).

The insistence on the interactive, dynamic properties of neural functions is in agreement with a systems theory approach. This approach is further developed in an eloquent statement of his position by Kinsbourne (1982):

> There are no discontinuities in the brain. No independent channels traverse it, nor is its territory divisible into areas that house autonomous processes. . . . No simulation of human behaviour, however impressively successful in impersonating its model, is capable of revealing how the human mind arrives at the same outcome, unless it is based on a network mechanism. (p. 412)

1.3.2 Localization of Function in Light of Systems Theory

It is possible to classify neuropsychological theories according to their position on the two dichotomies of localization versus non-localization and systemic versus nonsystemic. I will first describe the dichotomy between nonlocalization and localization views and then show how these views are modified by introducing the concept of systemic functioning.

1. *Nonlocalization, nonsystemic.* The theory says that functions are diffusely represented in a structurally relatively undifferentiated brain. Neural networks have been described as an alternative to localized functional centers. The alternative, however, seems hard to reconcile with the highly specific structures and patterns of connections found in modern neuroscience.

> The original idea of the neuron network as a continuum of nerve cells of standard shape and isotropic (random or geometrically determined) connectivity properties has all but disappeared from our image of the centers of the higher animals. (Szentagothai & Arbib, 1975, p. 43)

Therefore, because of the known specifity and diversity of anatomical structures in the brain, this theory cannot be maintained for the functioning of the brain as a whole. It may, however, be considered for certain functions in relation to limited brain regions. The thesis that lesions within the language areas give rise to aphasias of varying severity, but not of varying type, exemplifies this proposition.

2. *Localization, nonsystemic.* This theory states that the brain has highly specific and diversified anatomical structures with equally specific and diversified functions. The extreme example is phrenology,

which says that the brain is a collection of independently working organs.

The advantage of classical clinicopathological theory over phrenology is that it adds the possibility of integrative action by postulating connections between neural centers, as well as the building up of more complex functions by association. The theory is still nonsystemic, so long as the presence of connection does not modify the operations of localized functional centers. The Wernicke–Lichtheim model uses the concept of localized functions to describe and explain loss of function but adds aphasia syndromes (conduction aphasia, transcortical aphasias) caused by isolation or disconnection of language areas.

An even more sophisticated step in analysis is taken when localized centers are connected in temporal sequence and shifting combinations. These "functional systems" (see Luria, 1973) are seen as underlying normal performances. They are still not systemically organized in the sense of the present discussion, because the functioning of a component is not modified by other components, barring the special case of disconnection.

3. *Nonlocalization, systemic.* Although acknowledging the highly interconnected nature of neural tissue, I have already rejected the neural net as a sufficient model of the human nervous system. As an alternative to neural nets, Szentagothai and Arbib (1975) described more modern concepts, based on the idea of "modules" of neuronal organization. Although recognizing anatomical specificity, still the similarity of neuronal building blocks, called *modules*, throughout the cortex is stressed.

To make such a model systemic, it would be necessary to assume that the pattern of interaction between neural elements (modules) determines function, whereas this pattern can be set up anywhere in the brain, or at least in the cortex. Although such formulations are more theoretically acceptable than simpler concepts in the nonsystemic version of nonlocalization theory, the problem is to show how localized injury to a brain thus organized could result in a differential deficit in the language function, with relative sparing of other functions. Hence, the introduction of a systemic dimension does not make nonlocalization theories better able to explain the empirical findings. Although rejecting this alternative as a model for the brain as a whole, it may be that the language areas function in this way in relation to some language functions. In this model, small lesions may lead to slight or no defects, and lesions beyond a critical size may lead to a nonlinear increase of severity of deficit in several functions.

4. *Localization, systemic*. Whereas a nonsystemic localization theory holds that the function of a given area is determined by its anatomical structure alone, systemic localization would mean that the function of a given area is determined by its relationship to other areas within the bounds determined by its anatomical structure. This qualification is important in view of the well-known high degree of structural specificity of different anatomical regions in the brain. It may be, however, that given a ground plan of anatomical structure and connections, the pattern of function-to-structure relationship becomes more highly specific with time. If the ground plan is disrupted by injury, then some degree of rearrangement of the function-to-structure relationship may be possible. This flexibility in the development and the adjustment of the localization of function would have to be attributed to the interplay between the anatomical structures preserved at any given time. Such interplay must be mediated by neural pathways. Accepting the idea of a systemic element in the localization of function also entails a wider conception of the role of neural connection than that which is postulated in associationist schemes of brain function.

This model has the potential not only of explaining selective deficits with localized injuries, but also of explaining change or recovery of function in patients with structurally stable brain lesions. To my knowledge, no one yet has worked out a systemic localization theory of aphasia. The work of Luria (e.g., Luria, 1966), despite looking at symptoms or performances as based on functional systems, does not discuss systemic explanations of the phenomena of cerebral localization of function as such.

In a brain thus organized, the effects of lesions would be specific according to locus, but not additive with composite lesions. The effect of a lesion would be the result of a preinjury pattern of localization and the nature of the systemic response to injury in preserved tissue.

In summary, then, the selectivity of deficits with differently localized lesions is the best criterion for adopting a localization theory, whereas the additivity of effects in limited versus composite lesions is the main clue to the systemic organization of cerebral representation. In nonsystemic organization, the additivity of effects is preserved, but not in systemic organizations. Finally, the degree of changeability in the performance-to-structure relationship in recovery can be used as additional relevant information.

1.4. The Present Study

What one undertakes as a research project is determined by one's interests, practical limitations, and, most important, what one regards as reasonably well established.

During the period of time in which the material for this study was collected, the general services offered aphasic individuals in Norway were unsatisfactory. Speech therapy services were not organized and were available only in large cities or communities. No counseling, social support, or information pertaining specifically to the problems connected with aphasia were given to families, and no training in the care and treatment of aphasia patients was offered to hospital personnel. As a first step to improve this situation, the Institute for Aphasia and Stroke was established in 1973 by a donation from the National Health Association, a private organization with the fight against coronary and other vascular diseases as one of its goals.

The Institute for Aphasia and Stroke is a test laboratory located in the Sunnaas Rehabilitation Hospital, which is one of the municipal hospitals of Oslo, Norway. The staff of the institute consists of one neuropsychologist, one technician, and one research associate. The hospital has 226 beds and admits patients with several kinds of functional disturbance with organic etiology, offering physiotherapy, occupational therapy, speech therapy, and, in addition, social and medical services. Patients may be admitted for evaluation only, or for full treatment. The hospital offers only inpatient services; thus, the patients admitted for treatment have severe physical handicaps, whereas patients with lighter physical defects are referred to other institutions with outpatient services.

Because of its unique position, the institution received applications for admission from the entire country during the period of this study. An attempt was made to see as many as possible of the patients for evaluation and testing, so as to get a survey of the population referred for treatment. Although no exact figure can be given, it can safely be stated that more than 90% of the patients referred were tested.

The decision to create an aphasia registry was motivated by the desire for a systematic registration of all available information pertinent to the description and evaluation of the patient group. The registry should serve primarily as an instrument for clinical research, concentrating on the connection of aphasia with other symptoms and on the development of aphasia with time. The results of tests performed at the Institute for Aphasia and Stroke form the main content of the registry, with medical and general background information added.

As reviewed above, a summary of the consensus in 1978, when this study started, runs as follows:

The clinically defined syndromes of aphasia are stable entities with a well-defined pathological substrate. Because aphasia is a linguistic deficit, a more refined linguistic analysis of language performances in the major syndromes will allow us to replace the static traditional descriptions of functions as unanalyzed wholes with dynamic processing concepts approaching the ideal of complete computational specification with neural correlates.

My difficulties with accepting the position just summarized were based on both methodological and conceptual worries. First of all, I worried about the loosely defined procedures for testing and defining aphasic syndromes. A necessary first step for clinical research—and a step that must be taken anew in each different language community—is to define strict and quantifiable procedures for testing and classification. The system of myself and my colleagues is described in Chapters 2 and 3.

Second, I worried about the seemingly innocuous assumption that aphasia is a linguistic deficit. Remembering the papers by Teuber and Weinstein (1956), by Weinstein (1964), and by others showing an association of aphasia with some visual reasoning and learning tests, as well as the many exiting papers by Kimura (see Kimura, 1979) on the close association of language and higher order motor functions, I thought it more appropriate to define aphasia at the outset as a linguistic-cognitive defect. Although I in no way wish to question the reality of a language function separate from other cognitive functions, it may still be the case that aphasia does not reflect an isolated disturbance of this function. If it turns out that only a few selected cases demonstrate pure disturbances of language, then a scientific approach to the great majority is needed. Maybe these cases can be viewed as just "mixed" and can be explained as additions of defects observed in isolation in the pure cases. But it may also be that important interactions are at work, so that *mixed syndromes* becomes a misnomer for unanalyzed complexity.

Third, I worried about the generally simplistic approach to the effect of brain injuries evident in a deficit-oriented analysis. It seemed to me that the multidimensional response of the brain to injury, as well as the variations over time of this response, was what had to be described and accounted for. I was (and am) disturbed by the tendency to stress regularity and to dismiss variability in the response of the brain to injury as "noise." Commenting on the relative success of syndrome classification, Wernicke (1874) stated:

> Only a particular period in the course of the disease should be considered if one is to diagnose aphasia correctly. On the one hand, the general phenomena which accompany the onset of aphasia, as they do that of most localized lesions of the brain, must have disappeared. On the other hand, however, the conditions ought not to have lasted so long that the possibility of compensation by the other hemisphere is already present. (p. 69)

Here, the motivation of the neurological diagnostician to ignore information not pointing to the locus of the injury is clearly seen.

Poeck (1983b) echoed the same opinion:

> It cannot be denied that a certain number of vascular aphasias (approximately 15%) cannot be classified in terms of standard or nonstandard syndromes. The main reason, in our experience, is that the examination is done too early, prior to the establishment of a well defined syndrome, or at the late stage of recovery, with or without the effects of speech therapy. (p. 80)

On the basis of these reflections, and having standardized the necessary tools for measuring and classifying aphasic phenomena (Chapters 2 and 3), I have therefore undertaken a broadly conceived program of testing aphasics with neuropsychological tests. Information on lesions with CT-scans has been recorded when possible, and repeated testing has been performed in order to chart as far as possible the extremes of parametric values in aphasia, conceived of as an experiment of nature. I will attempt to give a systems-theory-oriented account of the complexity of the aphasic condition. It must be recognized that the account falls short of the goal of an experimental analysis of causally significant factors, if such an analysis can be given. It also falls short of the ideal of a specification of the actual processing stages behind the performances observed. I would still claim that, at the very least, this type of analysis of the organized complexity of linguistic-cognitive phenomena under a set of extreme conditions is a valuable complement to other sources of information about the underlying system.

OPERATIONALIZATION
OF A MODEL

2.1. The Model

The goal is to select a model that, to the best of our current knowledge, captures the significant dimensions of clinical syndromes. It should also account for the associations of parameters that are useful for defining syndromes while leaving reasonable space for within-syndrome variations.

The Wernicke–Lichtheim model underlies the terminology and the clinical classification systems most frequently used today. It is a model within the localizationist tradition (Wernicke, 1874; Lichtheim, 1885), and it identifies two cortical areas important to the language function, the Broca and Wernicke areas. The fibers associating these areas are assumed to run in the arcuate fasciculus. In addition, Lichtheim assumed that transcortical fibers, via a hypothetical "concept center," can mediate information between the language areas. Different forms of aphasia follow from lesions of different neurological structures.

The localization of language areas is shown in Figure 2.1, and the types of aphasia resulting from differently located lesions are shown in Table 2.1.

A more detailed review of brain regions and the associated aphasia types is given below. The review does not limit itself to statements

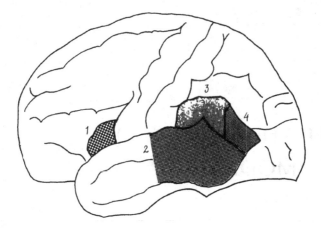

Figure 2.1. Localization of language areas. Legend: (1) anterior language area (Broca); (2) Wernicke area; (3) supramarginal gyrus; (4) angular gyrus.

by classical authors but intends to sketch the present-day status of this model.

2.1.1. Broca Area

The Broca area is located in the third transverse frontal convolution, which can be divided into three parts: the orbital, the triangular, and the opercular. It is the opercular part of the convolution

Table 2.1. Assumptions of the Wernicke–Lichtheim Model

Type of aphasia	Locus of lesion
Broca	Broca area
Wernicke	Wernicke area
Global	Broca and Wernicke areas
Conduction	Arcuate fasciculus
Anomic	Angular gyrus
Isolation syndrome	Extensive neocortical, sparing Broca and Wernicke areas
Transcortical motor	Frontal, sparing the Broca area
Transcortical sensory	Parieto-occipital, sparing the Wernicke area

that makes up the Broca center (Baily & von Bonin, 1951). This conclusion is corroborated by the results of electrical stimulation during
local anaesthesia in operations for epilepsy (Rasmussen & Milner,
1975).

The Broca area is designated Area 44 in Brodman's classification.
According to recent anatomical evidence summarized by Galaburda
(1982), Area 44 can be distinguished from surrounding cortex and
represents an intermediate degree of architectonic differentiation
between premotor cortex and primary motor cortex. Galaburda cited
evidence that interhemispheric asymmetries can be shown for parts
of the frontal operculum. The Broca area, like surrounding frontal
and lorbitall cortex, has evolved out of a proisocortical zone located
in the anterior insular region, and it maintains connections with this
more primitive zone.

Wernicke (1874) believed that the Broca area receives sensory
inputs from the musculature. It has the function of storing memory
("images") of performed movements. These images can be aroused
via association fibers from other cortical areas, thus giving rise to
speech. In the later literature, there has been recurring controversy
about the importance of the Broca area for language. The controversy
is at least partly conceptual. Some would assign the Broca area a
purely motor function and name the effect of a lesion of the Broca
area *anarthria* (Marie, 1906). Others would assign to it a special role
in the programming of speech movements but prefer to class the
resulting defect as a form of *apraxia* (Liepmann, 1915). Finally, some
would hold that the Broca area is essential for the activation of response
mechanisms in language, but that the deficits resulting from failure
can be distinguished from arthric and apraxic disturbances, and must
be properly classified as *aphasic* (Benson & Geschwind, 1977). Some
authors would deny that the Broca area has any function at all in
relation to language or speech (Pribram, 1971).

Broca aphasia is characterized by nonfluent speech, that is, speech
made up of poorly articulated short phrases produced with hesitations
and effort, particularly in initiation. Auditory comprehension is good,
but not completely normal. Ability to repeat and name is impaired,
but often better than the ability to produce words in spontaneous
speech. Reading comprehension is relatively good, but writing is
always impaired. (Goodglass & Kaplan, 1972; Benson & Geschwind,

1977; Mohr, 1976; Kerschensteiner, Poeck, Huber, Stachowiack, & Weniger, 1975).

2.1.2. Posterior Language Area

The posterior language area is composed of parts of the temporal neocortex, the gyrus supramarginalis, and the gyrus angularis. The *Wernicke area* is defined in this monograph as the temporal part of the posterior language area.

There seems to be a general agreement about the practical rule of thumb followed by neurosurgeons that the anterior part of the temporal lobe up to Labbé's vein can be excised without dire consequences for the language function. In neuroanatomical terms, Heschl's gyri are often given as the anterior limit of the Wernicke area. In regard to the posterior, it is generally agreed that this area is continuous with the supramarginal and angular gyri. The main disparity between diverse statements and diagrams seems to be that some regard only the superior temporal convolution as relevant to language, whereas others include the middle, and some authors even the inferior, temporal gyrus (Bogen & Bogen, 1976). The results based on electrical cortical stimulation in local anaesthesia vary among early reports (Penfield & Roberts, 1959), which seem to indicate a more extensive area, and later reports (Rasmussen & Milner, 1975; Fedio & van Buren, 1975), which find a more restricted area. I have adopted the definition that the Wernicke area consists of the posterior part of the superior and middle temporal gyri.

In the opinion of Galaburda (1982), the Wernicke area shows a degree of anatomic differentiation, judged by architectonic criteria, closely similar to that of the Broca area. He noted:

> In fact, architectonic similarities between anterior and posterior language areas and the overlap in their connectional organization make it a somewhat surprising finding that lesions in either region produce such different aphasic syndromes. (p. 443)

It should be noted, however, that the studies referred to have been performed on rhesus monkeys! The posterior language area has evolved out of proisocortex located in the temporal and posterior insular region. It contains regions of varying cytoarchitectonic differentiation, from primary auditory sensory cortex to more generalized neocortex, found in the inferior parietal lobule and the temporo-occipital junction.

Wernicke (1874) believed that the Wernicke area is a store of auditory word images (*Klangbilder*). The condition after injury is therefore characterized by difficulties with auditory language perception (total or partial word deafness) and disturbances of speech (because the appropriate auditory images for stimulating motor representations are disturbed).

Geschwind (1979) stated:

> Much new information has been added in the past 100 years, but the general principles Wernicke elaborated still seem valid. In this model the underlying structure of an utterance arises in Wernicke's area. It is then transferred through the arcuate fasciculus to Broca's area where it evokes a detailed and coordinated program for vocalization. (p. 187)

Wernicke aphasia is characterized by fluent, paraphasic speech and reduced auditory comprehension. Speech is produced without effort and has complex grammatical structure. Informational content is deficient (Goodglass & Kaplan, 1972; Huber, Stachowiack, Poeck, & Kerschensteiner, 1975). The term *jargon aphasia* is sometimes used for cases in which speech is totally incomprehensible, but jargon is not confined to Wernicke aphasia (Benson & Geschwind, 1977). Repetition is usually disturbed to the same degree as auditory comprehension, whereas naming performances may vary. Reading and writing are usually severely disturbed, but in some cases, they are preserved (Lecours & Rouillon, 1976).

The *supramarginal gyrus* is continuous with the superior temporal gyrus. Wernicke (1874) regarded the supramarginal gyrus as part of a continuous perisylvian gyrus, anatomically and functionally continuous with the superior temporal convolution. The stimulation data seem to support this view (Penfield & Roberts, 1959; Rasmussen & Milner, 1975). The reason for giving special consideration to this gyrus is that the probability of an auditory-language-comprehension defect is markedly lower with a lesion of the supramarginal gyrus alone than with a lesion of the superior temporal gyrus (Luria, 1970), whereas the probability of reduced fluency of speech with misarticulation and phoneme substitutions increases, possibly because of the proximity to the primary somatosensory cortex. The possibility of finding a syndrome corresponding to *conduction aphasia* may therefore exist with lesions of the supramarginal gyrus. Cytoarchitectonic studies by Galaburda, LeMay, Kemper, and Geschwind (1978) indicate that the part of the supramarginal gyrus immediately adjacent to the superior temporal gyrus may belong to the auditory association cortex, whereas more peripheral portions do not.

The *angular gyrus* is conventionally defined and is anatomically continuous with the middle temporal gyrus. According to Henschen (1922), only five cases with selective involvement of the angular gyrus had been published up to that time. The patients were all alexic and agraphic but had no auditory comprehension defect. Lesions of the angular gyrus occur often with more extensive involvement of the posterior language area. Authors who use an extended concept of the Wernicke area usually include the angular gyrus (e.g., Marie, 1906; Dejerine, 1914; Penfield & Roberts, 1959). Wernicke (1874) accorded no status to the angular gyrus in connection with reading and writing but assumed that association fibers from the occipital lobe to the Wernicke area were necessary for reading. The assumption of a special importance of the angular gyrus for reading and writing is widely adopted today. Some authors also accorded it a special function in word retrieval (naming) (Geschwind, 1967b; Luria, 1970). *Anomic aphasia* is characterized by fluent speech with marked shortage of content words. There is little paraphasia as such, but there are attempts to substitute circumlocutions and vague descriptions for content words. Comprehension and repetition are good, but there are severe problems in reading and writing (Goodglass & Kaplan, 1972; Kertesz, 1979).

A lesion encompassing both Broca and Wernicke areas produces a *global aphasia*. This may be regarded as a composite form of aphasia that should not be classified as an independent type, but clinically it has distinct features. It is characterized by severe loss in all language modalities, but the patient is usually not mute. Often, he or she has verbal stereotypes, consisting of conventional phrases (swearing) or meaningless syllabic combinations. In auditory comprehension, there is also some ability to react to concrete words, particularly if they are emotionally significant for the patient (Stachowiak, Huber, Kerschensteiner, Poeck, & Weniger, 1977).

2.1.3. Arcuate Fasciculus

Very precise descriptions of this cortico-cortical fiber bundle are hard to find. It has a compact middle portion sweeping around the insula parallel to the circular sulcus. The ends fan out and connect the inferior and middle frontal convolutions with large parts of the convexity of the temporal lobe. The existence of a direct projection from auditory areas to a homologue of the Broca area has been confirmed in rhesus monkeys by Pandya and Galaburda (1980).

This fiber bundle is one among several structures assumed to be of functional importance in connecting the posterior language area with the Broca area. Wernicke believed the insula to have this function, but this possibility is now considered unlikely, and in the neo classical literature, the arcuate fasciculus is accepted as the major functional connection. Lesions of the fascicle result in a *conduction aphasia* with relatively fluent speech and good comprehension, but repetition difficulty. Fluency may be less than in Wernicke aphasia, because the patient makes frequent attempts to correct literal paraphasias (phoneme substitution errors). He or she may go through a series of approximations in attempting to correct his or her production ("zeroing in"). In addition to good auditory comprehension, there is also often good reading comprehension (Benson, Sheremata, Bouchard, Segarra, Price, & Geschwind, 1973; Green & Howes, 1977; Benson & Geschwind, 1977).

Lesions outside the language areas mentioned above may produce aphasia. This means not that the areas injured have language functions, but that the language areas normally interact with surrounding areas when language is integrated in complex behaviors. The essential feature of such lesions producing aphasia is that they disconnect or isolate parts of the entire language areas from the surrounding cortex.

Total isolation produces an *isolated-speech-area syndrome*. The patient has no spontaneous speech but responds to questions. The response is almost always a direct repetition of the question. Speech is well articulated and the patient repeats even long sentences. He or she shows no sign of comprehension and fails in all language tests except repetition (Geschwind, Quadfasel, & Segarra, 1968).

Transcortical motor aphasia is characterized by an excellent ability to repeat and a sparse, but well-articulated, speech. Comprehension both of speech and of print is adequate, whereas writing is defective (Rubens, 1976).

Transcortical sensory aphasia has fluent, paraphasic speech and poor auditory comprehension. Unlike in Wernicke aphasia, however, the ability to repeat even long sentences is preserved. Reading and writing are usually severely defective (Kertesz, 1979).

2.2. Aphasia Test Construction and Standardization

In developing a suitable test methodology, the following considerations are important:

The test must cover the variables necessary and sufficient for classifying patients into types of aphasia in the Wernicke–Lichtheim version of a clinical pathological approach to aphasia.

The tasks selected for operationalizing these variables must be similar to the tasks used by other investigators in the same tradition.

A system of gradation must be developed, so that a comparison between individuals and between different performances within one individual is possible. This comparison presupposes that scores with acceptable statistical reliability are employed.

Some statistical justification must be given for grouping tests under a common heading. The scores must be checked for their sensitivity to extraneous variables, and appropriate corrections for such unwanted influences must be developed.

2.2.1. Test Variables

The variables necessary for classifying patients in the clinical-pathological tradition can be deduced directly from the description of aphasia syndromes by classical authors (Wernicke, 1874; Lichtheim, 1885). The necessary variables are the following:

Fluency of spontaneous speech
Auditory comprehension
Repetition

Naming
Reading comprehension
Reading aloud
Writing

Only the first four variables are critical to classification. The classical authors were uncertain about the frequency with which disturbances of reading and writing accompany the different aphasic syndromes and about the mechanism producing them. From the point of view of the controversy between "localizationists" and "holists," the inclusion of reading and writing among the basic variables affords an opportunity for comparing sense modalities (auditory and visual) and response modes (oral and graphic). All modern aphasia tests include tests of reading and writing.

In 1973, it was decided to construct a test battery comprising the above main variables for the purpose of classifying and grading aphasic disturbances. This work resulted in the publication of the Norsk Grunntest for Afasi (NGA) (Reinvang & Engvik, 1980b).

2.2.2. Selection of Tasks

A description of the type of tasks used to operationalize each variable follows. For details of the procedure, the Appendix to this volume must be consulted.

2.2.2.1. *Spontaneous Speech.* It is evaluated in response to specific questions ("What is your occupation?") and to open questions ("Tell me about your family"). The interview is tape recorded.

Three aspects of speech are evaluated:

1. Communicative function (0–4 rating)
2. Qualitative disturbance (0–3 rating)
   Literal paraphasia
   Complex paraphasia
   Visible effort
   Hesitations, pauses
   Stereotypy
   Dysarthria
   Self-correction
3. Quantity of speech

Words per minute (0–200)
Words per utterance (1,0–10,0)

Quantity of speech is scored on the basis of a transcript of the interview.

2.2.2.2. Auditory Comprehension. In the NGA, there are three classes of stimuli: objects, body parts, and language material. There are also three types of responses: pointing, complex acts, and yes–no choice. They are combined in the following tasks (the number of items is given in parentheses):

1. Body parts, point to the item named (1);
2. Body parts, point to the item described (6);
3. Body actions, carry out spoken instructions (10);
4. Objects, point to item named (11);
5. Objects, point to item described (6);
6. Objects, carry out spoken instructions (10);
7. Comprehension of ideas, respond yes or no (14);
8. Comprehension of relative statements, respond yes or no (4).

2.2.2.3. Repetition. The NGA includes the following types of tasks:

1. Word repetition (20)
2. Repetition of nonsense syllables (8)
3. Repetition of sentences (12)

Words are varied in terms of number of syllables, content (numbers and content words), and articulatory difficulty. These features are not scored separately. Nonsense syllables have been included as being representative of extremely low-probability words. They vary in number of syllables and stress pattern. Sentences vary in length (number of words) and content. They include examples of sentences loaded with function words ("Aldri annet enn om og men," meaning "Never anything but if's and or's") and of low-probability nonsense sentences ("Baaten sank i hytt og vaer," which can be translated as "The ship sank as the wind blows").

2.2.2.4. Naming. The material used in the auditory comprehension test is used again in the naming test. The following types of tasks are included:

1. Body parts, name (11);
2. Body actions, describe (5);

3. Objects, name (10);
4. Objects acted on, describe (5);
5. Responsive naming (10).

2.2.2.5. Reading and Sentence Construction. Printed stimuli are used, with letters, words, and sentences together with some objects from the test sample to investigate reading. Responses are oral (reading aloud), matching, pointing, or performing an action. Stimuli and responses are combined in the following tasks:

1. *Reading comprehension*, tested with recognition (6), word recognition (6), word–object matching (6), and printed instructions (5).
2. *Reading aloud*, tested with letters (6), words (10), and sentences (5 items, scored 2-1-0).
3. *Syntax*, sentence construction, tested with sentence fragments, to be arranged in correct order (6).

2.2.2.6. Writing. In the NGA, writing is tested with 10 items, including writing of the patient's own name, word copying, word dictation, written naming, and sentence dictation.

2.2.3. Similarity to Other Aphasia Tests

In comparing the methodology with other modern operationalizations of the same model, Benson and Geschwind (1977) are considered an authoritative source on recommended practice for clinical neurological investigation. Among modern aphasia test-batteries, reference is made to the Boston Diagnostic Aphasia Assessment (BDA) (Goodglass & Kaplan, 1972), the Western Aphasia Battery (WAB) (Kertesz & Poole, 1974; Kertesz, 1979), the Aachener Aphasietest (AAT) (Huber, Poeck, Weniger, & Willmes, 1983), and the Neurosensory Center Comprehensive Examination for Aphasia (NCCEA) (Benton, 1967). It may be doubtful whether the NCCEA is intended as an operationalization of the Wernicke–Lichtheim conception of aphasia. Benton (1967) gave as one of the purposes of the test

> To include specific tests in the battery which could be employed to investigate current questions in the field of aphasia (e.g. the reality of the clinical pictures of conduction aphasia, central aphasia and transcortical aphasia). (p. 41)

This statement must mean that the test comprises the necessary information for making a classification.

There is a high degree of overlap in the types of tasks included in these tests, and they are in significant agreement with clinical neurological recommendations of suitable tasks. The greatest variation is found in the procedure for evaluating fluency of speech. The NCCEA has no procedure for registering fluency. The BDA, the AAT, and the WAB use rating procedures, based on the same sort of qualitative observations as in the NGA. The NGA seems to be the first test to use quantified measures of speech (words per minute, utterance length) in clinical aphasia testing.

2.2.4. Choice of Normative Sample

The NGA has been standardized on a sample of 161 consecutive referrals with aphasia to a rehabilitation hospital (Reinvang & Engvik, 1980b). The reason for referral was mainly a request for treatment or for an evaluation of the indication for treatment. No study of other standardization samples has been performed.

In the Boston diagnostic test, the standardization was also limited to aphasics referred to the hospital (a VA hospital). It was felt that the selectivity of the group could be counterbalanced by developing separate norms for different severity-groups in the sample. This procedure was found, however, to affect only the level of the resulting profiles and not their form, and it was therefore dropped.

In the Western Aphasia Battery, control groups of nonneurological and neurological nonaphasic patients were used. For the groups with no involvement of the dominant hemisphere, the overall scores were above the 97th percentile for the aphasic group. In a group with diffuse involvement of the cerebrum, including the dominant hemisphere, but with no clinical diagnosis of aphasia, the score was at the 91st percentile for the aphasic group, but here one might question the criterion for the presence of aphasia in the control group. The results confirm the impression that nonaphasic individuals rarely fail the sort of items included in aphasia tests.

In connection with standardization of the NCCEA, Benton (1967) developed an elaborate procedure for age- and education-adjustment scores. He included a normal group in his standardization procedure but found that there was very little overlap between the aphasic and the normal populations. A reference group of aphasics was therefore used for developing norms for intra- and interindividual comparisons of aphasics. For the AAT, the main interest is to classify clinical syndromes, but as a part of standardization, nonaphasic controls were included in the same proportion as in clinical referrals. The tests, excluding speech rating, discriminated aphasics from normals with above 90% accuracy. Had spontaneous speech been included, the discrimination would have been near perfect (Willmes, Poeck, Weiniger, & Huber, 1980).

It may be concluded that the sort of tests described above, presumably including the NGA, discriminate poorly within the normal group. Altering the content of the tests to improve on discrimination among normals would lead to loss of discriminatory power among aphasics, as they would all be compressed into a narrow segment of the normal range or would fall totally below the normal range. A test constructed for the purpose of registering the type and degree of aphasic disturbances must therefore be standardized on a sample of aphasic patients.

The standardization sample is admittedly selective. One of the greatest clinical problems in connection with the clinical management of aphasia today is to find criteria for the selection of candidates for treatment and for differentiating the form of treatment among those candidates selected from the large population of individuals with unmistakable and often severe aphasic difficulties who seek such treatment. I believe that the intersection of selective pressures originating from the patient, his or her relatives, and the system of referral in Norway has produced a group that is representative in a limited sense, namely, representative of the type of aphasic who is currently considered a possible candidate for treatment. A test that is useful for establishing criteria for differentiation and prognosis in this group meets a significant clinical need.

2.2.5. Standardization

The standardization sample consisted of a subset of the total sample used in the rest of this study. It consisted of all aphasics admitted to Sunnaas hospital between 1974 and 1978, a period of about 4 years. This sample included 161 patients.

It was not regarded as necessary to restandardize the test for

Table 2.2. Mean, Standard Deviation, Range, and Reliability of Subtest Scores

Main variable subtest		N	Mean	SD	Range	reliab.
Communication		139	1.79	1.1	0–4	
Quality of speech	Literal paraphasia	132	.70	.9	0–3	
	Complex paraphasia	124	.56	.8	0–3	
	Visible effort	56	.55	.8	0–3	
	Hesitation	133	1.56	1.1	0–3	
	Stereotypy	131	.52	.9	0–3	
	Articulation	133	.86	1.0	0–3	
	Self-correction	51	1.12	1.1	0–3	
Fluency	Words per minute	70	48.89	39.7		
	Utterance length	72	3.92	2.4		
Auditory compre-hension	Body parts, identify	161	8.60	3.4	0–11	.93
	Body parts, describe	75	3.59	1.8	0–5	.87
	Body, actions	161	6.37	3.1	0–10	.87
	Objects, identify	73	9.19	3.0	0–11	.92
	Objects, describe	75	4.61	2.0	0–6	.89
	Objects, actions	161	6.70	3.2	0–10	.89
	Ideas, meaning	77	11.20	3.1	0–14	.84
	Ideas, relations	77	3.52	1.3	0–4	.66
	Total	73	53.4	17.5	9–71	.98
Repetition	Words	104	14.42	7.1	0–20	.97
	Nonsense syllables	106	4.96	3.0	0–8	.91
	Sentences	106	6.78	4.6	0–12	.95
	Total	104	26.10	14.1	0–40	.98
Naming	Body parts	161	6.59	4.4	0–11	.95
	Body actions	161	1.91	1.7	0–5	.89
	Objects	73	6.73	3.7	0–10	.94
	Objects, action	161	3.30	2.5	0–5	.90
	Responsive	161	6.12	4.2	0–10	.95
	Total	73	26.70	15.2	0–41	.98
Reading comprehension	Letters	161	5.04	1.7	0–6	.87
	Words	161	9.92	3.6	0–12	.95
	Sentences	161	3.19	2.1	0–5	.93
	Total	161	18.10	6.7	0–23	.97
Reading aloud	Letters	161	4.47	2.2	0–6	.91
	Words	75	6.43	3.7	0–10	.94
	Sentences	160	5.59	4.3	0–10	.95
	Total	75	16.90	9.6	0–26	.98
Syntax	Sentence arrangement	71	2.48	2.2	0–6	.83
Writing	Total	103	5.20	3.2	0–10	.88
Aphasia coefficient		68	150.80	61.1	14–217	.995

the total sample of the present study, as the test had been shown to have generally satisfactory statistical properties. To avoid unnecessary detail, the specific composition of the standardization sample is not shown. It contained 76% cerebrovascular cases, of whom two thirds were men. The mean age was 50. Further details can be found in the handbook for the test (Reinvang & Engvik, 1980b). A Danish edition of the test has appeared (Reinvang & Engvik, 1984), and a Swedish translation is used informally.

2.2.6. Statistical Properties

The performance of the standardization sample of subjects on the test is shown in Table 2.2 The subjects showed a wide range of performance, from almost no correct responses to perfect performance. The number of subjects varied somewhat because of revisions in the content of the test made during the time of standardization.

All the main variables (total scores) have a very high reliability measured by the alpha coefficient, a measure of internal consistency (Nunally, 1967). The test–retest correlations have been determined (Reinvang, 1981). They are reproduced here as Table 2.3 to confirm the impression of the high reliability levels given by the measure of internal consistency. The testing was done in a clinical context, which means that the tests and the retests were performed with long intervals. Recovery processes thus may have effected the results. The tests

Table 2.3. Test–Retest Correlations

	Acute	Chronic
Auditory comprehension	.93	.90
Repetition	.80	.94
Naming	.62	.93
Reading comprehension	.56	.89
Reading aloud	.70	.96
Syntax	.44[a]	.66
Writing	.58	.89
Aphasia coefficient	.82	.98

[a]Not statistically significant.

Table 2.4. Test–Retest Correlations

	Acute	Chronic
Communication	.87	.74
Literal paraphasia	.84	.68
Complex paraphasia	.40[a]	.54
Visible effort	.95	.87
Hestiation	.51	.75
Stereotypy	.94	.60
Articulation	.79	.80
Self-correction	.69[a]	.49
Words per minutes	.86	.95
Utterance length	.86	.98

[a]Not statistically significant.

are divided into acute (test and retest within 6 months after onset of aphasia) and chronic (later tests).

For the rating scales and quantitative measures of speech, no study of internal consistency could be made. In Table 2.4, the test–retest correlation coefficients have been given.

The correlations show some variability and are generally lower than the test–retest coefficients for objective scores. The quantitative measures (words per minute and utterance length) compare favorably with the rating scales. The reason that no study of intertester reliability was performed is that, at the time of standardization, very few persons except the author had been trained in administering the test.

The homogeneity of the main variables can be evaluated by inspecting the table of intercorrelations of subtests contributing to the same main variable (Table 2.5).

Principal component analyses with varimax rotation were performed to evaluate the loading of each subtest on the main variable to which it contributes. The loadings are generally very high and indicate that further splitting up of the main variables is not motivated by the standardization data.

The total scores have a high loading on a common factor (principal component analysis with varimax rotation), and this justifies the introduction of the sum of total scores on main variables, the aphasia coefficient, as a valid measure of the severity of aphasia (Table 2.6).

Table 2.5. Intercorrelations and Loading on First Factor for Subtests[a]

| | Intercorrelations | | | | | | | | Loading on 1st factor |
	1	2	3	4	5	6	7	8	
Auditory comprehension									
Body parts, identify									90
Body parts, describe	78								88
Body parts, action	86	81							91
Objects, identify	51	57	18						74
Objects, describe	71	73	68	83					88
Objects, action	79	74	82	70	83				92
Ideas, meaning	76	76	82	52	72	75			87
Ideas, relations	74	34	45	22	27	43	38		48
Repetition									
Words									96
Nonsense syllables	91								96
Sentences	86	85							94
Naming									
Body parts									96
Body parts, action	89								93
Objects	88	76							90
Objects, action	91	91	84						97
Responsive	90	85	84	92					95
Reading comprehension									
Letters									87
Words	64								90
Sentences	67	67							91
Reading aloud									
Letters									92
Words	80								95
Sentences	77	86							94

[a]Decimal points are omitted throughout the table.

Table 2.6. Intercorrelations and Factor Loading for the Main Variables[a]

Variable	Intercorrelations							Loading on 1st factor
	1	2	3	4	5	6	7	
Auditory comprehension								89
Repetition	68							82
Naming	85	80						91
Reading comprehension	84	57	75					88
Reading aloud	72	78	78	77				89
Syntax	57	47	61	63	62			75
Writing	65	58	62	69	64	65		80

[a]Decimal points are omitted throughout the table.

2.2.7. Relation to Background Variables

Studies with the NCCEA (Benton, 1967) have indicated that age corrections should be employed for some aphasia variables, and the work of McGlone (see McGlone, 1980) indicates that aphasia may be less severe in females than in males. No studies have been found, apart from Benton (1967), showing a relationship between the type of variables included in the aphasia test and education. In general, it is expected that the minimal overlap between the aphasia population and the normal population in language performance makes it unlikely that any strong relationship with education should be found. The relationship of aphasia test variables to age, sex, and education is summarized in Tables 2.7, 2.8, and 2.9. More detailed tables are given in the handbook for the test.

Differences were tested for significance by one-way analysis of variance and relationships at $p \leq .05$ have been given as significant.

There are few significant relationships between age and performance on different parts of the test. In general, the results indicate that separate norms should not be used for separate age groups. It must be added, however, that children were not represented in our standardization sample. The variables showing a weak interaction with age (repetition and naming) do not show a consistent trend of decreasing performance with increasing age. Hence, there is no reason to suggest systematic age-dependent adjustments of scores.

Table 2.7. Mean of Test Performance in Different Age Groups

	Age group						
Variable	19 and under	20–40	41–50	51–60	61–70	70+	*p*
Words per min.	62.8	31.7	45.4	52.4	61.8	45.0	n.s.
Utterance length	4.9	3.4	2.8	4.2	4.6	3.4	n.s.
Auditory comprehension	54.3	56.2	42.3	55.8	55.6	50.6	n.s.
Repetition	28.6	28.6	15.4	29.3	29.9	23.2	.05
Naming	29.7	29.2	12.8	31.1	28.1	27.4	.05
Reading comprehension	21.7	19.8	17.2	17.0	18.1	16.9	n.s.
Reading aloud	23.7	16.8	11.0	18.5	18.3	14.2	n.s.
Syntax	3.3	2.8	1.5	2.5	2.5	2.8	n.s.
Writing	5.3	5.9	4.7	4.9	4.8	6.2	n.s.

It may be concluded on the basis of Table 2.8 that there are few and unsystematic relationships between educational or professional level and performance on the aphasia test. The results indicate that separate norms for different educational groups are not motivated. In the one case of a significant relationship (reading comprehension), the tendency was for subjects with higher education to have poorer reading comprehension. This is probably an accidental finding.

The results in Table 2.9 indicate that there are few relationships between sex and performance on the aphasia test and that separate norms for males and females are not motivated. The observed difference on reading comprehension is probably an accidental finding.

In conclusion, the statistical studies of the standardization data indicate that, in this sample, the NGA measured aphasic performance with a high degree of reliability and consistency, and that performances showed very little dependence on age, sex, or educational level.

2.2.8. *System of Gradation*

In choosing a system of grading results, one finds that the two most likely systems are *z* scores and percentile values.

The *z* score is based on the assumption that the underlying

Table 2.8. Mean of Test Performance in Different Educational Groups

Variable	In school	Unskilled labor	Skilled labor	Artisan	High school + additional	College, university	p
Words per minute	61.7	58.4	46.0	42.1	54.5	61.5	n.s.
Utterance length	4.6	4.4	3.3	3.8	4.6	4.6	n.s.
Auditory comprehension	59.6	56.4	58.9	49.4	52.3	38.8	n.s.
Repetition	29.0	24.8	29.8	23.4	29.8	27.5	n.s.
Naming	33.4	28.9	31.9	22.9	26.0	19.8	n.s.
Reading comprehension	21.8	20.0	18.8	17.1	18.4	13.2	.05
Reading aloud	24.2	18.8	19.5	14.8	17.0	10.3	n.s.
Syntax	3.2	2.4	3.5	2.0	2.6	—	n.s.
Writing	6.1	4.6	5.7	4.8	5.2	5.6	n.s.

Table 2.9. Mean Test Performance in Males and Females

Variable	Male	Female	p
Words per minute	48.9	49.2	n.s.
Utterance length	3.9	4.1	n.s.
Auditory comprehension	52.5	54.9	n.s.
Repetition	25.5	27.3	n.s.
Naming	26.1	27.8	n.s.
Reading comprehension	17.4	19.7	.05
Reading aloud	16.5	17.5	n.s.
Syntax	2.5	2.5	n.s.
Writing	5.0	5.6	n.s.

from the mean in number of standard deviations, and the sign of the score tells if the deviation is positive or negative. The BDA assessment uses z scores to represent the results. These scores are advantageous for further statistical treatment but are open to the criticism that the underlying distribution of scores for most tasks used with aphasics is not normal.

Percentile values tell what proportion of scores falls above or below a given value. If a raw score of 10 correct responses corresponds to a percentile value of 25, that means that 25% of the standardization

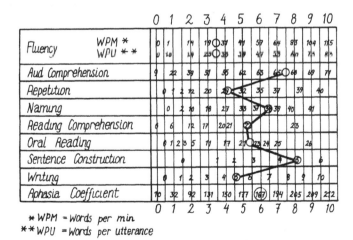

* WPM = Words per min.
** WPU = Words per utterance

Figure 2.2. Example of aphasia test result.

population scores below this value and 75% score above. Percentile values are less suitable than z scores for various statistical treatments but have the advantage of not presupposing any form of the underlying distribution. Percentile values are used in the NCCEA and in the WAB.

In the NGA, percentile values are used to represent scores. The main usage of percentile values is to make individual judgments in the form of test profiles, whereas for statistical studies on groups the untransformed raw scores are used.

An example of a test result is given in Figure 2.2. The raw scores are circled, and the scale indicates that, for example, a raw score of 38 on naming is at the 65th percentile of the distribution of aphasics.

How to proceed from the test profile to the determination of aphasia type is described in the next chapter.

TYPES OF APHASIA

The nomenclature and criteria for types of aphasia in this study are based on the Wernicke–Lichtheim model, as described in Chapter 1. Within this tradition, different test batteries have different rules or guidelines for determining the aphasia type after converting raw scores to derived scores and examining the resulting test profiles.

The Boston Diagnostic Aphasia Assessment (Goodglass & Kaplan, 1972) gives only guidelines for interpretation together with an indication of the range of variability in scores within a given type of aphasia. Clinical judgment is recognized in addition to scores as a valid basis for classification.

The Western Aphasia Battery (WAB) gives more strict quantitative definitions with exact cutoff values.

The classification system and criteria are shown in Table 3.1.

The scores are percentage scores, except for fluency, which is rated directly on a 10-point scale.

In the system based on the Aachen Aphasia Test (AAT), clinical judges have classified a reference group of aphasics. A computer program decides the likelihood that new patients will be assigned to any subcategory of the reference group, and a probability of at least 80% is necessary to accept a classification (Willmes et al., 1980).

In choosing between a system with a strict quantitative criteria and one with room for clinical judgment, one can argue that a study that intends to explore the relationship of aphasia type to locus of lesion and to neurological signs must strive for maximally specific definitions of aphasia type, as otherwise the risk of the judgment's

Table 3.1. Aphasia Classification System of WAB

		Criteria for classification		
Aphasia	Fluency	Comprehen-sion	Repeti-tion	Naming
Global	0–4	0–3.9	0–4.9	0–6
Broca's	0–4	4–10	0–7.9	0–3
Isolation	0–4	0–3.9	5–10	0–6
Transcortical motor	0–4	4–10	8–10	0–8
Wernicke's	5–10	0–6.9	0–7.9	0–9
Transcortical sensory	5–10	0–6.9	8–10	0–9
Conduction	5–10	7–10	0–6.9	0–9
Anomic	5–10	7–10	7–10	0–9

Note. From *Aphasia and associated disorders: Taxonomy, localization and recovery* (p. 58) by A. Kertesz, 1979, New York: Grune & Stratton. Copyright 1979 by Grune & Stratton, Inc. Reprinted by permission.

being contaminated by information on the dependent variable would be present (i.e., the fallacy of psychoneurology). For research purposes, a strictly quantified system therefore seems necessary.

A classification system like that in the WAB has the property of being exhaustive; that is, any patient with a complete aphasia-test result is assigned to one of the groups in Table 3.1. It may be argued that a classification system based on a clinicopathological model should allow for unclassifiable or "mixed" cases. This conclusion seems a natural consequence of choosing a model that stresses selective deficits in patients with limited lesions. In clinical practice, some patients must have composite lesions, and hence mixed symptomatology. The proportion of such cases depends on the selectivity of action of the agent of injury, as well as the selectivity of the sample. For research purposes, it may be justified to exclude unclassified cases from a study because they tend to obscure the findings. Clinically, however, they are common and must be characterized by some diagnostic term.

Both in the Boston diagnostic test and the Western Aphasia Battery, the rules for classification are *a priori*, that is, based on arbitrary definitions of cutoff values. It would be possible to study empirical classifications based on purely statistical criteria so that optimal cutoff points for predicting a given outcome (e.g., locus of lesion) could be established. This procedure is followed in the AAT, where the criterion variable is a clinical classification. Another way to derive an empirical classification is to study clusterings of scores within a sample of aphasia test performances by advanced statistical methods. In previous research by Kertesz (1979), both the methods of *a priori* quantitative definition and of empirical, statistically derived classification have been used. On the basis of these studies, it seems that a likely development is a further refinement within the framework of a classical clinicopathological classification system. A clustering analysis performed on the AAT by Willmes *et al.* (1980) gives groups that overlap extensively with the clinical classifications.

3.1. Classification System of the Norsk Grunntest for Afasi

In the Norsk Grunntest for Afasi (NGA), strict, quantitative criteria for division into types are suggested, so that, given a test profile, the type designation follows automatically. The cutoff values chosen for this study have been determined by the author's experience and judgment. The classification adopted by the NGA is not exhaustive; hence, unclassifiable cases occur. No patient or test is excluded from the study because of "mixed" findings or other peculiarities of the test result. The definitions of aphasia types proposed here differ from those used by Kertesz *et al.* mainly in being relational. Rather than focusing on the absolute level of, for example, comprehension in Broca aphasia, it is stressed that comprehension must be better than fluency and that the difference must exceed a certain cutoff score. This type of rule is intended to allow a characteristic configuration of performances to be designated by a given name although the level of performance might improve. Relational definitions are most appropriate with Broca, Wernicke, conduction, anomic, and transcortical aphasias, where clinical descriptions all note relational features. All aphasia types should not be defined relationally, however. In global aphasia, the uniform severity of the deficit across performances should be stressed. The term *global aphasia* refers only to patients with nonfluent speech. I have chosen to include jargon aphasia as a separate type, and to define it as uniformly severe aphasia but with mixed or fluent speech. By having separate terms for all the most severe aphasics, regardless of fluency, confounding of type and severity of aphasia may be reduced in further analyses comparing performance in different types of aphasia.

3.1.1. Definitions of Speech Classification

If a tape recording exists so that quantitative evaluations can be made, then the rules are the following:

Nonfluent speech: Words per minute below 40 and utterance length below 4.0.

Fluent speech: Words per minute above 80, utterance length above 5.0 and presence of paraphasia (literal or complex).

The cutoff values follow the values suggested by Kerschensteiner, Poeck, and Brunner (1972). The cutoff point for the upper limit of nonfluency is also close to the 50th percentile in the empirical distribution (see Chapter 2).

Intermediate: The criteria for neither nonfluent nor fluent speech are satisfied, and speech is neither normal nor predominantly dysarthric.

Dysarthric: Fluency cannot be determined because of a strong dysarthric component in the speech. Dysarthria does not preclude that there is also an aphasic disturbance.

Normal: Speech is quantitatively in the fluent range, but without presence of paraphasia or other qualitative signs.

If a tape recording is not available, then nonfluent speech is diagnosed on the basis of a high rating on hesitation, visible effort, and stereotypy. Fluent speech means the presence of complex paraphasia or literal paraphasia, with a relative absence of visible effort or hesitation. If one class of ratings does not clearly predominate over the other, then fluency is mixed.

3.1.2. Definitions of Aphasia Types

The most characteristic features of different types of aphasia are outlined in Table 3.2, whereas precise, quantitative definitions are given in the following sections.

3.1.2.1. Types of Aphasia with Intermediate or Fluent Speech. Anomic

Table 3.2 Characteristics of Different Aphasia Types

Fluency	Auditory comprehension	Repetition	Naming	Aphasia type
Fluent	High[a]	High	Low	Anomic
		Low	—	Conduction
	Low	High		Transcortical sensory
		Low	Low	Jargon
		Low	—	Wernicke
	Others	—	—	Mixed, fluent
Intermediate	High	Low	—	Conduction
	Low	Low	Low	Jargon
	Others	—	—	Mixed, with mixed fluency
Nonfluent	High	High	—	Transcortical motor
	Low	High	Low	Isolated speech area syndrome
		Low	Low	Global
	Others	—	—	Mixed nonfluent

Note. For precise definitions, see text.
[a]The terms *high* and *low* may be defined conditionally.

aphasia has fluent speech. Naming is more than 20 percentile points worse than repetition and fluency.

Conduction aphasia has intermediate or fluent speech and comprehension is more than 20 percentile points better than repetition.

Transcortical sensory aphasia has fluent speech, and repetition is more than 20 percentile points better than auditory comprehension.

Jargon aphasia has fluent or intermediate speech. Auditory comprehension, repetition, and naming are below the 20th percentile. The aphasia coefficient is below the 20th percentile.

Wernicke aphasia has fluent speech, and auditory comprehension and repetition are more than 20 percentile points below fluency. The aphasia coefficient is above the 20th percentile.

Mixed aphasia with fluent speech is the group of remaining patients with fluent speech; they have test profiles not classifiable as jargon, Wernicke, transcortical sensory, anomic, or conduction aphasia. This is not regarded as a type of aphasia in the classical literature.

Mixed aphasia with intermediate speech is the group of patients in whom spontaneous speech cannot be classified as either fluent or nonfluent, and the test profile is not classifiable as jargon aphasia or conduction aphasia. This group is not homogenic and must be regarded as unclassifiable cases.

3.1.2.2. *Types of Aphasia with Nonfluent Speech.* Transcortical motor aphasia has nonfluent speech, and auditory comprehension and repetition are more than 20 percentile points better than fluency.

Broca aphasia has nonfluent speech and auditory comprehension more than 20 percentile points better than fluency. Naming is more than 20 percentile points better than fluency.

Isolated-speech-area syndrome has nonfluent speech and repetition more than 50 percentile points better than auditory comprehension and naming.

Global aphasia has nonfluent speech and scores below the 20th percentile for auditory comprehension, repetition, and naming. The aphasia coefficient (AC) is below the 20th percentile.

Mixed aphasia with nonfluent speech is not a type of aphasia in conventional classifications. The group consists of the remaining patients with nonfluent speech who cannot be classified as having global, isolation syndrome, Broca, or transcortical motor aphasia.

Table 3.3. Characteristics of the Sample

Characteristic	Mean	Median	SD	Range
Aphasia coefficient (AC)	129	135	62.6	1–216
Age at test (in years)	50.2	53.1	15.9	11–80
Time from illness to test (in days)	260	134	372	2–2,131

3.2 Subjects

The classification rules and other empirical questions were tested on a main sample described in the following (Table 3.3). It consisted of 249 patients included on the basis that a complete aphasia test and scores on key neuropsychological variables were recorded. Patients with predominantly dysarthric speech were excluded.

There were 161 men and 88 women in the sample. In comparison with the standardization sample (Chapter 2), it may be noted that the aphasia as measured with AC was on the average more severe in the main sample. There were 84% cerebrovascular patients, the large majority of thromboembolic origin. Head injuries were represented by 11% in the sample and miscellaneous other diagnoses by 5%.

The results of applying the classifications rules for aphasia types to this sample are shown (Tables 3.4 and 3.5). When more than one test was performed, the first was used for classification.

3.3 Stability of Classification

When more than one test has been done, the stability of the classification can be studied. This is not strictly a reliability measure, as the influences of the recovery process are at work when the time span between tests is several months. Still, under these conditions, it should be pointed out that the test–retest coefficients of the

Table 3.4. Speech Fluency

Nonfluent	129	(52%)
Fluent	34	(14%)
Intermediate	78	(31%)
Normal	8	(3%)

Table 3.5. Aphasia Type

Global	37	(15%)
Isolation syndrome	2	(1%)
Transcortical motor	3	(1%)
Broca	19	(8%)
Jargon	12	(5%)
Wernicke	11	(4%)
Transcortical sensory	6	(2%)
Anomic	8	(3%)
Conduction	23	(9%)
Nonfluent, unclassifiable	67	(27%)
Others	61	(24%)

quantitative test variables are very high (see Tables 2.3 and 2.4; the tables on the stability of classification are Tables 8.2 and 8.3).

Significant stability is present in a statistical sense; that is, the probability of being classified in a syndrome is not independent of previous classification. Still, the probability of reclassification is quite sizable (36%), and this seems somewhat undesirable. Rather than having a complex set of intersyndrome movements with improvement, it would be desirable that a patient retain his or her classification. If improvement involves highly specific patterns of recovery, however, then syndrome reclassification may be accepted as a genuine finding.

3.4. Comparison with Typology of WAB

With respect to the WAB, the similarity in test procedure is high enough so that a comparison of outcomes can be made. On the basis of the classification criteria of Kertesz, a group of patients tested with the NGA can be classified according to the criteria used in WAB. The relationship of the two classification systems is shown in Table 3.6 (from Sundet & Engvik, 1984).

The NGA is more restrictive in assigning a classification, but of the patients who are classified, 85% are given the same classification by the WAB. Some of the apparent disagreement is spurious. The WAB does not use jargon aphasia as a separate category. Given this premise, it is reasonable that the 8 patients called jargon asphasics in the NGA should be classified as Wernicke aphasics in the WAB. In all, it is reasonable to regard patients with a given syndrome

Table 3.6. Cross-Classification of Cases in Two Systems

Norwegian Basic Aphasia Assessment	Western Aphasia Battery								
	Global	Broca's	Isolation	Trans. mot.	Wernicke's	Trans. sens.	Conduction	Anomic	
Global	23								28 (15%)
Broca's	1	20		2					23 (12%)
Isolation			1						1 (1%)
Trans. mot.	1			2					3 (2%)
Jargon					8				8 (4%)
Wernicke's					6		1		7 (4%)
Trans. sens.					2	1			3 (2%)
Conduction					5		12	2	19 (10%)
Anomic								3	3 (2%)
Mixed, nonfluent	29	21	3	1	1				55 (28%)
Mixed, mixed		3		2	16	4	2	12	39 (20%)
Mixed, fluent						1	1	2	4 (2%)
	59 (31%)	44 (23%)	4 (2%)	7 (4%)	38 (20%)	6 (3%)	16 (8%)	19 (10%)	193

Note. From The validity of aphasic subtypes by K. Sundet and H. Engvik, 1984, June. Paper presented at INS-European Conference, Aachan, West Germany. Reprinted by permission of the authors.

designation in the NGA system as a subset of patients with the same diagnosis in the WAB system.

Sundet and Engvik (1984) also performed a cluster analysis after the guidelines suggested by Kertesz and Phipps (1977) and described the 9 most prominent clusters (Table 3.7). The corresponding table from Kertesz and Phipps (1977, Table 2) is reproduced as Table 3.8.

In interpreting the clusters, it is apparent that Clusters 1 and 2 comprise severe aphasias with nonfluent speech, whereas 4 and 5 represent less severe nonfluent patients, including some whose speech rate placed them in a borderline zone.

Cluster 3 clearly captures severe aphasias with fluent speech output, whereas 6, 7, and 8 have the less severe fluent patients, with the qualification that Clusters 7 and 8 have a high proportion of cases with less-than-fluent but not nonfluent speech.

3.5. Levels of Classification

On the basis of these quantitative analyses one might question the desirability of a fine-grained classification system for all purposes.

Table 3.7. Cluster Composition of Patients in the NGA System

Cluster	Number of patients	Percentage of aphasia types
1	25	96% global
2	20	15% global 85% mixed nonfluent
3	16	50% jargon 25% Wernicke
4	19	74% mixed nonfluent
5	32	31% Broca 63% mixed nonfluent
6	3	66% conduction 33% anomic
7	29	31% conduction 52% mixed intermediate
8	19	32% conduction 11% anomic 53% mixed intermediate
9	23	57% Broca 9% trans. mot. 22% mixed intermediate

Table 3.8. Cluster Composition of Patients in the WAB System

Cluster	Number of patients	Percentage of clinical aphasia types
I	30	97% global
II	15	86% Broca's
III	12	25% global 25% Broca's 25% isolation
IV	13	54% Broca's 23% isolation 23% transcortical motor
V	4	100% transcortical sensory
VI	12	58% conduction
VII	11	100% Wernicke's
VIII	7	57% conduction
IX	18	63% anomic
X	20	100% anomic

Note. From "Numerical taxonomy of aphasia" by A. Kertez and J. B. Phipps, 1977, *Brain and Language, 4*, pp. 1–10. Copyright 1977 by Academic Press, Inc. Reprinted by permission.

Rather than saying, "Aphasics should be grouped in this way, because that is how they are grouped in nature," it seems possible to adopt both a fine-grained and a coarse-grained classification system depending on the objective.

It is questionable if a syndrome division as derived from the Wernicke–Lichtheim model is fruitful for all purposes. The empirical results so far suggest that a fourfold division of aphasias into mild versus severe deficit and nonfluent versus relatively fluent speech captures the main divisions of the empirical structure. This seems to be the view also of the Aachen group, who operate with four main syndromes: global, Broca, Wernicke, and anomic aphasia. Poeck (1983b) stated:

> Aphasic syndromes are expressive syndromes. They do not have a clearly defined receptive aspect, although language comprehension is compromised in all aphasic patients, even though to a different degree. (p. 85)

Comprehension is, next to fluency, the most important criterion for classification in the Wernicke–Lichtheim model.

A fourfold classification system has also been used in many

reports from the Milan group (Basso, Vignolo, and others; see Basso, Capitani, & Vignolo, 1979; Basso, Capitani, Luzzati, & Spinnler, 1981).

The possible contribution of time to classification must be acknowledged. Even at the level of coarse-grained analysis, some distinction between acute and chronic symptomatology should be made.

A coarse-grained classification system can be used, then, when general structures underlying performance are analyzed with psychometric methods. For the analyses presented in this study, a classification system with three dimensions is used: fluency, severity, and chronicity. Each of these variables is dichotomized so as to prepare the ground for psychometric analyses with ANOVA designs and opportunities for measuring interactions.

The cutoff point for fluency is set so that all nonfluent patients form one group. The cutoff point for severity is at the median value of the aphasia coefficient of the sample. For classifying chronicity, I follow the same practice as Kertesz and Phipps (1980), regarding aphasia as chronic when 6 months or more have elapsed since onset. This is somewhat longer than the median of the sample.

The resulting structure of the sample is shown as a quasi-three-dimensional figure (Figure 3.1).

3.6. Sample Structure

An analysis of the relation of the classification system to sex, age, etiology, and education was performed to ascertain the presence of any systematic bias in the groups. The relationship was measured with chi-square tests when the dependent variable was discrete (sex,

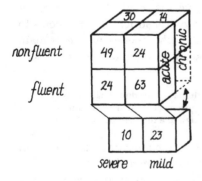

Figure 3.1. Sample structure.

education, and diagnosis) and with ANOVA for age. Only in the latter analysis could interactions be tested, and these were found to be insignificant (see Table 3.9).

The age variable was weakly related to fluency, and the finding was that nonfluent patients were younger than fluent patients (48.8 vs. 50.0 years). More importantly, the severe aphasics were older than the mild aphasics (52.2 vs. 46.6 years). The one significant finding on education is probably not important: twelve subjects were still in school, and of these, 10 had a mild degree of aphasia.

With diagnosis, the relation to fluency was just marginally significant ($p = .05$). Traumatic patients who tended to fall in the fluent group account for the trend. More important is the relationship to severity, which is that traumatic patients tended to fall in the mild group.

The tendencies support earlier reports that head-injured patients have a higher probability of being fluent than vascular cases. This finding has been reported for closed head injuries by Heilman, Safran, and Geschwind (1971).

There is also a tendency for fluent aphasics to be older than nonfluent aphasics, as reported by Obler, Albert, Goodglass, and Benson (1978). This tendency is also present in the series reported by Kertesz and Sheppard (1981), who reviewed alternative explanations of the finding. It may be related to the changing organization of the aging brain, as suggested by Brown and Jaffe (1975); to different survival probabilities for patients with different aphasia syndromes; or to differences in the etiology of cerebrovascular disease with age, as suggested by Kertesz and Sheppard (1981).

The conclusion must be that, in analyses showing a difference of severe versus mild aphasics, the age bias must be corrected for, unless the variable in question is known to be uncorrelated with age. For some analyses the separation of vascular from nonvascular groups should be considered also.

Table 3.9. Summary of Relation of Aphasia Classification to Background Variables

	Age	Sex	Education	Diagnosis
Nonfluent vs. fluent	*	—	—	*
Severe vs. mild	**	—	*	*
Acute vs. chronic	—	—	—	—

Note:— = no significant difference; * = significant difference at p < .05; ** = significant difference at p < .01.

3.7. Conclusion

In summary, a simplified classification system based on the dimensions fluency, severity, and chronicity is justified by empirical findings and offers advantages for the purpose of psychometric studies. This system is exhaustive and classifies all patients. The reality of selective aphasia syndromes, some of them rare, is confirmed. The criteria applied here are consistent with, but stricter than, the criteria used in another major operationalization of the Wernicke–Lichtheim model, that of Kertesz (1979). The system is exclusive and did not classify 51% of the cases in this sample. The possible advantage was that the resulting groups were homogeneous. Traditionally, the relation of classification to locus of lesion has been the center of interest. The strongest argument for traditional typology is that it is claimed to predict lesion localization. As noted in Chapter 1, even the critics of a clinical-pathological model or a localizationist approach have accepted that the traditional typology has a predictive value. The value of the present well-defined, but restrictive, system remains to be seen (see Chapter 7).

More recently, the use of clinically defined types for neurolinguistically oriented analyses of language subprocesses has been debated. The current opinion goes in the direction that only studies of single cases or of very homogeneous small samples are of value in such studies (see Schwartz, 1984). These questions cannot easily be resolved, and for the time being, we must live with different classification systems, differing both in type and in degree of inclusiveness, and must evaluate their usefulness in relation to the objective of study.

SELECTIVE APHASIAS

4.1. Types of Selective Aphasias

Under this heading is usually discussed a varied set of conditions that have the common feature of only affecting verbal performance in a given modality (auditory or visual, oral or graphic) while the general level of language functioning is unimpaired. The latter condition is very rarely fulfilled since at least a mild aphasic disturbance is usually present. The existence of deficits specific for comprehension or naming of some categories of words but not of others have also been reported.

4.1.1. Modality-Specific Aphasias

The types of modality-specific aphasias discussed in the literature include *agraphia* (selective disturbance of writing), *alexia* (selective disturbance of reading), *aphemia* (selective disturbance of speech), and *auditory verbal agnosia* (pure word deafness), meaning selective disturbance of auditory language comprehension. In these cases, it can be shown that whatever sensorimotor disturbances are present are not sufficient to account for the verbal deficits, because other patients with the same sensorimotor deficits do not have the same verbal deficits. All these forms of aphasia are rare.

In aphemia, the patient has severe speech difficulty both spontaneously and on imitation. The presence of intact swallowing and tongue motility distinguishes the condition clearly from anarthria, although the patient may show clumsiness and inaccuracy in imitating oral movement. Articulation is impaired even for automatized sequences (e.g., counting). In the recovery phase, the patient speaks slowly and with effort but typically produces full sentences with no sign of agrammatism. In the conceptual scheme of Wernicke–Lichtheim, this form of aphasia is termed *subcortical motor aphasia*, with the implication that the lesion disconnects output from the Broca area from motor effector organs.

Pure agraphia is a very rare syndrome but has been described by Exner (1881); more recent studies have been summarized by Laine and Marttila (1981). Some of these cases have evident problems with the motor aspects of writing and should be viewed as cases of an apraxic nature. Some patients, however, have symptoms that are more unequivocally of a language nature; one such case was described in detail by Bub and Kertesz (1982). This patient produced *semantic paragraphias*, writing concrete names from dictation but not being able to write pronounceable pseudowords. No consistent locus of lesion has been found for patients with pure agraphia.

Pure alexia (alexia without agraphia) has been studied in a series of cases starting with Déjerine (1892). Geschwind and Fusillo (1966) gave impetus to renewed interest in the syndrome, which they explained as a disconnection syndrome (i.e., visual input is cut off from the language areas). Patients with pure alexia have lesions of the posterior left hemisphere, almost invariably resulting in right-side hemianopsia. They typically read isolated letters or numerals with some success but fail to read words or, indeed, to identify letters in a word context. In addition to hemianopsia, inability to name colors (*color anomia*) is frequently present. Patients may show varying degrees of *visual agnosia* (i.e., a general inability to identify or name visual stimuli). Although the disconnection explanation of pure alexia has been widely accepted, opposition to it has been voiced by Hécaen and Kremin (1976) and Levine and Calvanio (1982), both of which groups have found evidence of a left-hemisphere mechanism for the identification of letter groups. The destruction of this mechanism, rather than the disconnection of right-hemisphere visual input from left-hemisphere language processes, is involved in pure alexia.

In addition to pure alexia (alexia without agraphia), *aphasic alexia* (alexia with agraphia) may be noted as a semipure defect. In these cases, speech and auditory comprehension is relatively intact, whereas disturbance of reading and writing may be severe. Several patterns may be exhibited. The relative preservation of letter reading with impaired word reading has been mentioned.

In the syndrome of *deep dyslexia* (see Coltheart, Patterson, & Marshall, 1980), the patient produces semantic paralexias, reads isolated letters and meaningless letter combinations poorly, and reads concrete nouns with fair success. Hécaen and Kremin (1976) found sentence alexia to be a distinct variety of alexia.

Auditory-verbal agnosia (pure word deafness) is a rare condition found in some cases of bilateral temporal-lobe pathology with signs of more general auditory agnosic defects (Ulrich, 1978). A case without pathological verification, but very likely only left-hemisphere involvement, was reported by Gazzaniga, Velletri Glass, Sarno, and Posner (1973). In Wernicke aphasia, recovered cases may report the experience that the speech of others sounds like a foreign language; as one patient said, "I can hear, but the sound doesn't come all the way through." These cases may demonstrate relatively intact reading comprehension and thus may approximate the condition of auditory verbal agnosia.

4.1.2. Material-Specific Aphasias

There have been a few reports of deficits specific to certain categories of words. Goodglass, Klein, Carey, and Jones (1966), comparing data for comprehension and production (naming), made the observation that the hierarchy of difficulty for categories of material varies strongly between the conditions. Goodglass and Kaplan (1972) remarked on the frequent occurrence of comprehension failure for names of body parts. Dennis (1976) found a specific inability to produce or understand body part names in a young girl after resection of the left anterior temporal lobe. The author concluded that the patient's problem could not be explained by a disconnection of sound and meaning but involved lexical selection within the category of body parts.

Color-naming defects are found primarily in connection with symptoms of dyslexia or visual agnosia (de Renzi & Spinnler, 1967; Kinsbourne & Warrington, 1964; Geschwind & Fusillo, 1966). The suggestion that there is a second form of color anomia occurring in the context of aphasia and lesion of the language areas (Oxbury, Oxbury, & Humphrey, 1969) has not won much support.

Geschwind (1967b), in his classical paper on naming errors, considered the hypothesis that the usually observed generalized naming impairment is an agglomeration of category-specific naming defects. He pronounced the issue to be of great theoretical importance but found no empirical basis for resolving it. Poeck and his associates (Orgass, Poeck, & Kerschensteiner, 1974; Poeck & Stachowiack, 1975) rejected the notion of category-specific naming or comprehension deficits. They pointed out that, in unselected aphasic patients, comprehension and naming of different materials are highly correlated (on the order of .6–.7). This is obviously not a telling argument because lesion size may well account for correlations of that magnitude. Specific color-naming defects are found only in patients with callosal syndromes. These researchers concluded that the more interesting differences between aphasics are in the use and comprehension of linguistically defined categories (e.g., word class), and this has been the dominant research trend for the last 10 years.

4.1.3. Models of Selective Aphasias

Although rare in themselves, the pure aphasias are interesting for the light they may throw on the mechanisms for integrating sensorimotor and language processes. Historical discussions centered on the question of whether all the "pure" cases could be explained by a disconnection of the language areas from input or output. The alternative was to assume separate "centers" for acquired language-dependent skills, and proposals were made for locating centers for both writing and reading. The second frontal convolution and the angular gyrus were the respective candidates. This thinking originated in a strict associationistic and hierarchic processing model in which all perceptual processes take place bottom-up (from sensory to symbolic levels) and all the motor control processes in top-down fashion.

Recent neuropsychological analyses of dyslexia and agraphia have led to models emphasizing parallel and alternative coding systems for graphic material. Assume that reading can be accomplished both by a letter-to-sound translation process and by a whole-word-to-meaning matching procedure. The selective impairment of one of these routes for reading predicts essential features of the known dyslexia syndromes rather well. The two-routes-of-reading hypothesis as an explanatory model for dyslexia was proposed by Marshall and Newcombe (1973), and the analysis was taken up and extended in a later book (Coltheart et al., 1980).

In the case of writing, Friederici, Schoenle, and Goodglass (1981) proposed two independent encoding systems at the word level, one phoneme-to-grapheme conversion system and one word-to-graphic-pattern conversion system. In line with this proposal, Bub and Kertesz (1982) described the syndrome of deep agraphia in an analogy to deep dyslexia, which may not be explained as failure of the phoneme-to-grapheme conversion mechanism.

What these models do is to reject the hypothesis that sensory systems interact with symbolic systems (language) only at very low levels of the linguistic code. This conclusion may have wide implications, and before discussing it further, I examine some data on normals.

4.2. Studies of Sensory Mechanisms and Language in Normals

The main interest of psycholinguists and cognitive psychologists has been to show the influence of higher levels of information processes on lower level perceptual processes. In the well-known click experiments, Fodor and Bever (1965) showed that the perceived temporal location of a click presented during a sentence was influenced by the syntactic structure of the sentence.

In the proofreading situation, it is commonly assumed that attending to the semantic content of the text makes it more difficult to detect misprints and spelling errors. The literature on memory in an information-processing context usually makes the assumption that sensory-based memory (preserving the sensory qualities of the stimulus) is of short duration and that more durable memory is based on verbal-symbolic recoding in which information on physical characteristics is lost. (More on memory in the next chapter.)

If higher level processing interacts with perceptual processes, there is no *a priori* reason that the interaction may not be bidirectional. In that case, the notion of hierarchically organized levels becomes harder to maintain.

An initial indication that linguistic information is coded with reference to sense modality comes from experiments on the so-called priming effect. The effect is that words that have been presented before are easier to recognize than unfamiliar words. It has been found that, in order to facilitate visual recognition, the repeated words have to be presented visually. Increased familiarity on the basis of auditory presentation does not facilitate visual recognition. This finding led Morton (1979) to postulate separate visual and auditory word-recognition units in his logogen model. Further evidence for modality-specific learning in the processing of text comes from studies by Kolers and coworkers (Kolers, Palef, & Stelmach, 1980) and from Levy (1983). The latter found that familiarity with a text improves the ability to detect printing errors. The improvement takes place only in the case where familiarity is acquired by reading the same text in the same script. Neither scrambled words in the same script, the same text in a different script, nor the same text presented auditorily leads to any facilitation in detecting printing errors.

Hasher and Zachs (1979) suggested that some aspects of stimuli are encoded automatically. Automatic encoding is characterized by an occurrence independent of variations in age or type of instruction and by a minimum of effort. There is some evidence that frequency of occurrence and spatiotemporal context are automatically encoded stimulus aspects. The work on priming effects referred to above suggests that sensory modality may also be automatically encoded. Lehman (1982) studied recall for word lists given in mixed auditory or visual presentation. There was uniformly high recall for the modality of presentation regardless of age and whether the testing for recall of modality was expected or unexpected. Instructions ensured that the words were processed for meaning. Even the subjects' recalling the words themselves by grouping them taxonomically across modalities did not interfere with recall of modality. The author concluded that the modality of presentation is automatically encoded in long-term memory, and that the importance of modality in the organization and retrieval of information is unknown.

Finally, some evidence of the mutual independence of the processes underlying reading and writing in normals may be cited. Spelke, Hirst, and Neisser (1976) found that subjects could be trained to read a text for meaning while writing from dictation, without interference between the two activities.

All these studies indicate that the representation of sensory and linguistic information is highly specific and integrated. The problem of the activation and retrieval of linguistic information based on input into a sensory modality may not best be seen as that of connecting two separate representations but as that of coordinating aspects of a unitary representation. *Unitary* in this context means that the linguistic and nonlinguistic codes are integrated. Codes are *multiple* or distributed in the sense that the same "chunk" of linguistic information may be coded in different nonlinguistic contexts.

4.3. The Present Study

The occurrence of selective (material- or modality-specific) aphasias in the present patient sample was studied.

For material-specific deficits, the question of the selective impairment of body part comprehension and naming was studied. The relevant parts of the aphasia test were analyzed with principal factor analysis to determine if a material-specific factor was present. If so, the results were analyzed further with relation to aphasia classification. *Statistical Package for the Social Sciences* (SPSS) (Nie, Hadlai Hull, Jenkins, Steinbrenner, & Brent, 1975) was used for performing the analyses.

For modality-specific deficits, factor analysis is not a suitable method. As in the classification of types of aphasia, an *a priori* system was used, based on differences between performances exceeding a predetermined cutoff point.

4.3.1. Material-Specific Deficits

The relevant parts of the aphasia test are summarized in Table 4.1.

A factor analysis was performed on the results of these tests and factors with eigenvalues above 1.0 were rotated with the varimax

Table 4.1. Selected Aphasia Subtests

	Material	
	Body parts	Objects
Auditory comprehension	Identify from name (BP–C1) Identify from description (BP–C2) Perform action (BP–C3)	Identify from name (O–C1) Identify from description (O–C2) Perform action (O-C3)
Naming	Confrontation naming (BP–N1) Naming of action (BP–N2)	Confrontation naming (O–N1) Naming of actions (O–N2)

procedure to determine the loadings of the different tests. The hypothesis was that a factor with loadings from all the tests referring to body parts could be found.

The factor analysis yielded two factors accounting for 83% of the variance. The loading of the subtests on these factors is given in Table 4.2. It is evident that the two factors separated naming and comprehension rather than body parts and objects. The hypothesis was thus not confirmed.

Table 4.2. Factor Analysis of Tests with Body Parts and Objects

	Factor 1	Factor 2
BP–C1	—	.78
BP–C2	—	.70
BP–C3	—	.70
BP–N1	.88	—
BP–N2	.91	—
O–C1	—	.84
O–C2	—	.84
O–C3	—	.76
O–N1	.87	—
O–N2	.91	—

4.3.2. Modality-Specific Deficits

The tests used in the assessment of reading and writing are summarized in Table 4.3. The predetermined criteria for diagnosing a modality-specific defect used a minimum difference of 20 percentile points as a cutoff score. The criteria were of the same type as those used in the definitions of aphasia type (Chapter 3). The definitions and frequencies for the different categories are given in Table 4.4.

The breakdown with respect to type and severity is shown in Table 4.5. The association of each selective deficit with the type and severity of aphasia cannot be tested with a statistical method giving a measure of interaction. The association with severity is less interesting because the opportunity for detecting selective deficits is not present in very severe cases (floor effect). Therefore, chi-square tests of association were performed on the association of nonfluent or fluent speech with alexia, agraphia, auditory verbal agnosia, and hypergraphia. No significant associations were found.

Bearing the weak statistical basis in mind, I offer some comments on the apparent trends in the data. Alexia and agraphia were found more frequently in mild, fluent cases, a finding consistent with traditional notions of an association of these deficits with anomic aphasia (see Benson & Geschwind, 1977). The trend was more marked for agraphia than for alexia, reminding us that alexia may also be found

Table 4.3. Tests of Reading and Writing

Reading comprehension	Point to letters	(6 items)
	Point to words	(6 items)
	Match word to object	(6 items)
	Perform instruction	(5 items)
Reading aloud	Name letters	(6 items)
	Say object names	(6 items)
	Say polysyllabic abstract words	(4 items)
	Say sentences	(5 items)
Writing	Own name	(2 items)
	Copy words	(2 items)
	Word dictation	(2 items)
	Object naming	(2 items)
	Sentence dictation	(2 items)

Table 4.4. Definition and Incidence of Modality-Specific Deficits

Type	Definition	N	%
Alexia without agraphia	Reading compreh. < aud. comprehension[a] Reading aloud < repetition and naming	15	(6)
Alexia with agraphia		4	(2)
Agraphia	Writing < repetition and naming	16	(6)
Auditory verbal agnosia	Reading compreh. > aud. comprehension	17	(7)
Hyperlexia (pure)	Reading aloud > repetition and naming	10	(4)
Hypergraphia (pure)	Writing > repetition and naming	31	(12)
Hyperlexia with hypergraphia		9	(4)
		102	(41)

[a]Difference exceeding 20 percentile points.

Table 4.5. Relation of Modality-Specific Deficits to Type and Severity of Aphasia

	Severe nonfluent	Mild nonfluent	Severe fluent	Mild fluent
Alexia without agraphia	5	2	1	7
Alexia with agraphia	0	1	0	3
Agraphia	2	3	1	10
Auditory verbal agnosia	3	2	4	8
Hyperlexia	1	2	4	3
Hypergraphia	13	4	5	9
Hyperlexia with hypergraphia	1	3	2	3
Total	25	17	17	43

in patients with nonfluent speech (Benson, 1977). Hypergraphia was found in all groups, but the occurrence of about 40% of the cases in the severe nonfluent group is interesting. One report (Mohr, Sidman, Stoddard, Leicester, & Rosenberger, 1973) emphasized dissociation of oral and written responses in global aphasia. Auditory verbal agnosia was found more frequently in the fluent group. Localization of lesions is discussed in Chapter 7.

4.3.3. Conclusion

One likely reason for these divergent findings can be found in the literature reviewed initially. Impairments of more than one mechanism can cause disturbances of function, and statistical criteria as used here cannot distinguish between disturbances caused by different mechanisms. This is an area in which experimental single-case studies have been valuable, and the present approach must limit itself to a descriptive outline.

The findings do not support a notion of material-specific comprehension or naming mechanisms for body parts. Modality-specific deficits defined by a relative criterion were found in all major groups of aphasia. This result would argue against viewing these deficits as a product of only one mechanism, disconnection of the language areas from input or output. It is not denied that this mechanism exists. It is, however, also necessary to take into account the integrated nature of the sensory and linguistic representations and the possibility of selective deficits of this integrated code.

MEMORY AND LEARNING
DEFICITS

5.1. Normal Memory

What kinds of memory are there? According to Tulving (1982), there are at least three kinds: procedural, episodic, and semantic. *Procedural memory* is concerned with the performance of skills and corresponds to "knowing how." *Episodic memory* is memory for events in a spatiotemporal context, whereas *semantic memory* is conceptual knowledge. The last two forms of memory are, in principle, different from the first because the question of veridicality can meaningfully be asked about them. You believe that you saw John in the stolen car, but is it really true? You know that all cars have wheels, but is it really so?

Memory may be of a short-lived or a more permanent kind. The type of encoding processes performed on stimuli are important in determining their memorability. The distinction between sensory memory (iconic or echoic), short-term memory, and long-term memory served originally to delineate memory systems with different decay functions (Sperling, 1960; Atkinson & Shiffrin, 1968). Major coding and recoding operations are the transfer from sensory to short-term memory by phonetic articulatory recoding (Conrad, 1964) and the transfer from short-term to long-term memory by rehearsal (Atkinson & Shiffrin, 1968). This by-now-classic theory places great emphasis on verbal-articulatory mechanisms in coding. It has been modified in several ways. Sensory memory that is not verbally recoded is not necessarily short-lived. Nonverbal codes have their own long-term representations (see Crowder, 1976, Chapter 3). Even verbal information may enter long-term or semantic memory without going by way of phonemic recoding, as discussed in connection with reading in Chapter 4. A notion of processing memory or working memory has also gained access to memory theories. It is a memory mechanism holding information from different memory stores plus information on the current environment with a view to making decisions. It is fair to say that memory theories have become less strictly hierarchical and less inclined to view memory as exclusively stimulus-driven (see Broadbent, 1984, and the ensuing discussion for a review).

There are no pure tests for memory components, but different experimental paradigms have been developed to study memory. Different task analyses have been proposed for these paradigms, and many experimental variations have been used to elucidate these hypotheses. For convenience, paradigms may be subsumed under the headings of *memory* and *learning tasks* without implying that mechanisms responsible for performance respect these boundaries.

5.1.1. Verbal Memory

Immediate memory is measured in clinical practice by having the patient repeat a string of digits, letters, or words and noting the patient's success with different list lengths. This procedure tells something about the capacity for storage, but not about the duration of memory. Reproduction after variable delays may be introduced to study information decay, and to prevent rehearsal in the delay interval, a distractor task can be used. Immediate memory does not directly reflect short-term memory. Reproducing six or seven digits probably involves more than one memory mechanism. The initial items have probably been coded in a long-term memory storage, whereas the last item(s) is coded in short-term memory (STM), the very last item even in precategorical acoustic storage (Crowder & Morton, 1969). Glanzer (1972) argued for this multiple-memory-systems explanation of the position effects in verbal free-recall paradigms, where both the initial and the final items are better recalled than those in the middle positions.

5.1.2. Verbal Learning

The paradigms frequently used are the learning of word lists and paired associate learning. Tulving (1972) pointed out that these paradigms are not good models for the acquisition of genuinely new knowledge. Known words with their category membership and associations are used, and hence, what the subject is mainly required to learn is the particular pattern of co-occurrence of items. Although stimulus-and-response difficulty may also be manipulated as experimental variables with significant effects, a component of episodic memory requirement is thus present. Further evidence for factors other than the use of semantic or conceptual knowledge comes from studies of serial learning, in which the stimulus sequence is fixed. Ebenholz (1972) has shown that the serial position of items is coded and facilitates the learning of a new list in which some old items are presented in the same serial position as compared with different serial positions.

5.1.3. The Relation of Verbal Memory and Learning to Language Function

There is no consensus that memory processes are impaired in the syndrome of amnesia (see Hirst, 1982). These patients have problems recalling day-to-day events and learning new material when learning requires several trials. Skill learning is preserved, and the patients profit from retrieval cues. Episodic memory is believed to be involved, whereas semantic memory is largely intact. Amnesic patients are not regarded as having language defects. People with poor conceptual knowledge or bizarre beliefs may be regarded as stupid, ignorant, or insane, but they are not usually characterized as language deficient.

Another variation of clinical memory defects is the rare cases with very limited auditory verbal immediate memory described by Warrington and Shallice (1969) and Basso, Spinnler, Vallar, and Zanobio (1982). The injuries producing such deficits usually involve the language areas, but the patients are only mildly aphasic.

There are three possible hypotheses about the relation of verbal memory to aphasia:

1. A verbal memory deficit causes the language impairment or certain aspects of it.
2. The language impairment causes deficient memory because encoding into the verbal code is impaired. Such encoding is important for durable memory.
3. Verbal memory impairments exist independently of language impairment.

The first hypothesis is unlikely in view of the above remarks on clinical memory disturbances and language. The second hypothesis seems almost self-evident in view of the importance of verbal coding processes in theories of memory, but the memory mechanisms in on-line language processing may only partly overlap with those tested in memory experiments. The third hypothesis therefore has a fair chance of success.

5.2. Verbal Memory and Learning in Aphasics

5.2.1. Verbal Memory

Aphasics have poor immediate memory as assessed by their ability to repeat digits or letters, pointing to digits or letters spoken or shown, pointing to pictures corresponding to spoken words, or recognizing words previously heard (Goodglass, Gleason, & Hyde, 1970; de Renzi & Nichelli, 1975; Cermak & Moreines, 1976). To account for these deficits, some authors have considered the second hypothesis above.

The experimental investigation of rehearsal effects was started by Heilman, Scholes, and Watson (1976), who hypothesized that the inability to repeat of Broca and conduction aphasics was the cause of their poor performance on tasks of immediate memory. The absence of an interference effect between acoustically similar stimuli had previously been noted by Goodglass, Denes, and Calderon (1974). This was taken to indicate the nonuse of covert verbal mediation by aphasics in a memory task that involves such mediation in normals. Rothi and Hutchinson (1981) used a paradigm that allowed an independent assessment of immediate memory (repetition) and informational decay over filled and unfilled intervals. Of all groups studied, nonfluent aphasics showed the poorest immediate reproduction, but no loss of information with distracting tasks (counting). Fluent aphasics, on the other hand, showed somewhat better (but subnormal) immediate reproduction and an interference effect of the distracting task. This finding leads to the conclusion that all aphasics show a deficit in verbal memory, but fluent aphasics make some efficient use of a verbal rehearsal mechanism, whereas nonfluent aphasics do not. As nonfluent aphasics demonstrate significant retention, the question of which memory mechanisms they use is an intriguing one. Rothi and Hutchinson (1981) discussed the possibility of direct semantic coding on the basis of sensory information, the slowness of this process accounting for the information loss.

The conclusion seems to be that a defect in verbal rehearsal leading to an inefficient "working memory" (Baddeley & Hitch, 1974) is present in nonfluent aphasics, but that it does not account for all findings.

In exploring the third hypothesis, that different memory impairments may coexist with and interact with aphasia, the possibility of a memory impairment peculiar to nonfluent aphasics has already been stated (they do not have the type of memory with the decay characteristics of STM).

The presence of an acoustically based memory defect is central in the interpretation by Luria (see Luria, 1973) of the syndrome of Wernicke aphasia. In his terminology, it is an "acoustico-mnestic aphasia," and the basis of it is an auditory perceptual defect not specific to verbal material. In connection with studies of auditory-verbal short-term memory deficits, Shallice and Warrington (1974) found normal memory for environmental sounds and concluded that the deficit studied by them was modality- and material-specific. Gordon (1983) found decrement in auditory immediate recall for both digits and tonal stimuli in aphasics with lesions of the Heschl's, superior, and middle temporal gyri, as well as the inferior parietal lobule. Both auditory-verbal short-term memory deficit (Warrington, Logue, & Pratt, 1971; Warrington & Shallice, 1969) and selective auditory-perceptual language deficit (pure word deafness) (Gazzaniga et al., 1973) can exist independently of aphasia, but it would not be surprising if any one of these deficits could be associated with aphasia.

The presence of a recency effect may give clues about which is the most common type in association with Wernicke aphasia.

Specific impairment of auditory precategorical storage (PAS) (Crowder & Morton, 1969) is thought to explain the recency effect in verbal learning, the effect being that items in the final position of auditorily presented lists are recalled better than items in middle or prefinal positions. When verbal material is presented visually, the effect is not present.

The literature on sequential errors refers to studies by Efron (1963) and Swisher and Hirsh (1972). These have shown that, in order to reliably judge the order of two successively presented auditory stimuli, the aphasic patient requires a time separation one order of magnitude greater than the normal subject. Moreover, it is patients with posterior lesions who show this deficit phenomenon to the most extreme degree. The deficit in perceptual ordering of nonlanguage auditory signals shows no direct correlation with auditory language comprehension. Tzortzis and Albert (1974) extended the study of ordering deficits to language material. Three conduction aphasics showed retention for content but not for order of words in a short-term memory task. The authors suggested that this deficit underlies the repetition deficit in conduction aphasia. The study by Heilman, Scholes, and Watson (1976) opposed this conclusion and found no qualitative and quantitative differences in memory scores between Broca and conduction aphasics.

The failure to encode temporal characteristics of the stimulus sequence may be relevant to the form of memory referred to as *episodic* and may be different from the encoding of stimulus content.

In conclusion, it seems that aphasics show distinctive verbal memory deficits that may be specific to the type of aphasia, although the findings are by no means conclusive.

5.2.2. Verbal Learning

Given that aphasic patients have poor short-term verbal memory this does not totally determine their capacity for verbal learning. Current theories of memory admit of direct coding into long-term memory without short-term storage, and even with limited short-term memory, the possibility of alternative learning strategies may cause differences in the efficiency of learning.

There is a conceptual and methodological difficulty in interpreting deficiencies in verbal learning as memory difficulties. It is likely that aphasia leads to some alterations in the structure of the premorbid lexicon (Zurif & Caramazza, 1976). Because new information interacts with already-coded information in long-term storage, it is difficult to pinpoint the source of a deviance resulting from this interaction. It is reasonable to describe the findings without making strong theoretical claims.

Carson, Carson, and Tikofsky (1968) described results for aphasics and controls with common verbal learning paradigms including serial learning. They concluded that the aphasics showed normal learning curves, but with a generally lower level of achievement than the controls. Howes and Geschwind (1964) described two groups of aphasics, Types A and B, of which Type B show disturbances in word associations. It is natural to assume that they have disturbances relating to semantic coding that interfere with learning that requires semantic grouping or association. The prediction was apparently confirmed by Beauvois and Lhermitte (1975), who measured immediate memory span for words and the learning of eight-word lists. Patients with semantic paraphasia showed normal immediate memory and severe learning impairment. Patients with exclusively phonemic paraphasia showed reduced immediate memory span but normal learning.

5.3. Nonverbal Memory and Learning

Tests of nonverbal memory and learning have been used in research that contrasts patients with left- and right-hemisphere pathology. A series of studies from the Montreal Neurological Institute have shown the resulting impairments to be material-specific; that is, patients with left-temporal-lobe pathology showed deficits with verbal but not with nonverbal material, and patients with right-temporal-lobe pathology showed the reverse pattern. These patients were only minimally aphasic or nonaphasic (see Milner, 1974).

De Renzi and Nichelli (1975) used some of the methods developed in the above research to study a broader range of patients with left- and right-hemisphere pathology. Using the Corsi block-tapping test (see description in Section 5.5.1), a nonverbal analogue of digit span, they found that patients with right-hemisphere lesions were inferior to patients with left-hemisphere lesions, who, in turn, were inferior to controls. The presence, type, or degree of aphasia was not important in explaining the results. The authors are inclined to believe that the deficit in the left-hemisphere group was related to a disturbance of attentional factors present mainly in patients with posterior lesions. De Renzi, Faglioni, and Previdi (1977) extended the previous research by including a learning task with block-pointing sequences of supraspan length. Again, it was the patients with visual field defects who did poorly, especially those with a right-hemisphere lesion. The presence or the type of aphasia was not reported in this study. The authors reported another study in which a subspan sequence of three blocks was reproduced after filled or unfilled intervals of 6 or 18 seconds. All groups showed some loss of information even without interference, but the loss was increased with a verbal interference task (counting). The pattern was the same in controls and in patients with right- and left-hemisphere injuries.

Cermak and Tarlow (1978) tested memory for words, pictures of objects, and nonsense shapes in a continuous recognition paradigm. Although severely impaired for words, the aphasics showed normal memory for pictures. The nonsense shape task proved too difficult for even the control group. It was concluded that the memory deficit showed by the aphasics was material-specific, and not related to any attentional or perceptual difficulty with the stimuli. The aphasics all had nonfluent speech.

5.4. Conclusion

The above discussion leaves many unresolved questions. First of all, there are the questions of the specificity of the memory disorders for verbal material. For short-term or immediate memory tasks, it has been well documented that specific deficits exist that are neither a direct correlate of the severity of the aphasia nor an aspect of a generalized memory disorder.

To the extent that some of the memory defects reported in aphasia refer to the spatiotemporal encoding of verbal sequences, there seems to be no reason why they should be specific to verbal material. It is consistent with a material-specificity hypothesis that, as long as tasks are matched for difficulty, the encoding of the nonverbal aspects of the task or the use of response strategies should not be specific to material type. If the material-specificity hypothesis is unfounded, then the alternative hypothesis is that task types are organized as integrated wholes cerebrally, and not composed of constituents that can be varied independently. The learning tasks may serve to throw light on this problem.

Second, there are the questions of subtypes. Above it has been hinted that at least two types of short-term-memory deficits may exist, one material-specific and another modality-specific. To the extent that these are based, in the one case, on a close association with defective rehearsal and, in the other, with defective precategorical acoustic storage, a study of serial position curves may be useful in dissociating them. Research has focused on the association of short-term-memory defect with specific forms of aphasia, especially with conduction aphasia. In view of the pervasive difficulties that aphasics have with such tasks, it does not seem likely that the difficulties are restricted to specific aphasia types, although this question is still open. The careful definition of type as independent of severity of aphasia seems crucial in this context.

5.5. Present Study

The study compares the results of different procedures and paradigms, described in more detail below, to throw light on the questions raised about the material specificity and the nature of verbal memory and learning problems in aphasics. A distinctive feature of the approach was the attempt to apply parallel tasks of a verbal and nonverbal nature to the study of these problems.

The subjects for the study are the population of 249 patients described in Chapter 2. For tasks demanding verbal performance, it was reasonable to exclude patients with insufficient verbal ability to comply with the instructions, and their exclusion led to a loss of cases in different analyses. To avoid unnecessary detail, this exclusion is not discussed for every test, but the requirements for inclusion are stated; for the most demanding tests, 35% to 40% of the patients were excluded.

First, the tests used are described and discussed with regard to findings in previous studies with aphasics and with control groups. Then, the structure of the memory functions underlying the performance on the tests is analyzed with the method of factor analysis. Last, the relation of memory performance to aphasia groups is studied. The format adopted for analysis is the cube model described in Chapter 2, with ANOVA performed on the three main dimensions (type, severity, and chronicity) and testing for two- and three-way interactions (SPSS; Nie *et al.*, 1975).

5.5.1. Tests

A summary of the tests used and the scores and measures derived from them is given in Table 5.1.

5.5.1.1. Verbal Immediate Memory

5.5.1.1.1. Digit Span. The digit span test is the repetition in the same order of a series of digits presented orally. The procedure for administration and scoring is as in the WAIS-test (Wechsler, 1958). The results were not taken as an indication of memory, unless the patient could repeat one digit correctly.

The normal performance on span forward and backward was, on the average, 6+5 according to the Norwegian standardization (Engvik, Hjerkin, & Seim, 1980). Costa (1975) found the mean span forward and backward to be 5.2 and 3.3 for left-brain-damaged patients and 5.5 and 3.4 for right-brain-damaged patients. In left-lesion non-aphasic patients, de Renzi and Nichelli (1975) reported a mean span of 5.7 forward. Black and Strub (1978) divided their left-lesion group into patients with frontal and posterior lesions and found 5.8 forward and 4.2 backward for frontals and 4.9 and 4.1 for posteriors. The group included 8% aphasics. The question of whether digit span forward and digit span backward measure the same underlying function has been discussed. Rudel and Denckla (1974) suggested that

Table 5.1. *Summary Table of Tests Used*[a]

Function	Test	Derived measures (code)	Range
Verbal immediate memory	Digit span	Digit span forward (DF)	1–8
		Digit span backward (DB)	1–8
	Pointing span	— (PS)	1–18
Verbal learning	Digit serial learning	Trials (DS-T)	1–17
		Error type: Perseveration (DSE-P)	0–1
		Sequence (DSE-S)	0–1
		Intrusion (DSE-I)	0–1
		Refusal (DSE-R)	0–1
		Position score: Initial (DSP-I)	0–2
		Middle (DSP-M)	0–2
		Last (DSP-L)	0–2
			0–2
	Verbal association	Easy-item score (WPA-A)	0–18
		Difficult-item score (WPA-B)	0–9
Nonverbal immediate memory	Block pointing	Block span forward (BF)	1–10
		Block span backward (BB)	1–10
Nonverbal learning	Block serial learning	Trials (BS-T)	1–17
		Error type: Perservation (BSE-P)	0–1
		Sequence (BSE-S)	0–1
		Intrusion (BSE-I)	0–1
		Refusal (BSE-R)	0–1
		Position score: Initial (BSP-I)	0–2
		Middle (BSP-M)	0–2
		Last (BSP-L)	0–2
	Shape association	Trials (SA-T)	0–20

[a]For description, see text.

span backward is partly determined by visuospatial abilities. Costa (1975) found some confirmation of this, but Richardson (1977) and Black and Strub (1978) did not.

   5.5.1.1.2. *Pointing Span.* Pointing span was constructed after the procedure used in a study by Goodglass *et al.* (1970). The test materials were two displays consisting of cardboard plates with six objects pictured on each. The subject was instructed to point to the pictures as they were named by the examiner. If she or he failed more than once on each display, the test was discontinued, and the score was not used as an indication of memory function. If the criterion was met, the test was continued by asking the patient to point to series of pictures of up to five items in the order named by the examiner. The test items were prerecorded on tape and were played back through a loudspeaker from a Tandberg 3000 X tape recorder. The display with pictures was covered while the patient listened to the tape. Removing the cover was the signal for the patient to start pointing. One point was awarded for the correct performance of a task, and the results for two displays were summed for a total score.

5.5.1.2. *Verbal Serial Learning.* After digit span forward had been established, a new series of digits was presented containing two digits more than the estimated span. The order of presentation remained constant. Repetition of this series continued until a criterion of perfect recall on two consecutive trials was reached. Responses after each trial were noted. The score was the number of trials needed to reach criterion. Hamsher, Benton, and Digre (1980) found that normal subjects learned sequences of eight digits without problems even in groups of high age. For patients with a span of five digits or less (up to seven digits in the learning task), a scoring of performance on the first, middle, and last digit was performed on Trials 1 and 2. They were scored as correct when present regardless of ordering. For patients who had not mastered the task in seven trials (the mean score), the error type most prevalent was scored. The predetermined alternatives were perseveration, intrusion of erroneous items, faulty sequence, and refusal to continue. The responses on each trial were recorded, and judgment on error type was performed by the author on the basis of the protocols. A patient was permitted to score on, at most, two error types. Ratings were given as presence or absence of the given type.

5.5.1.3. *Verbal Associative Learning.* A *Wechsler paired-associate learning* test, easy and difficult items, was taken from the Wechsler memory scales (Wechsler, 1945). No Norwegian standardization is available, so the author's translation was used. A series of 10 word pairs were read three times. After each presentation, recall was tested by the examiner's saying the stimulus word and the subject's attempting to recall the response. The order of presentation was changed between reading and recall of the list, and between each reading of the list. The 10 word pairs contained 6 easy pairs (*up-down, north-south, metal-iron, baby-cries, fruit-apple, rose-flower*) and 4 difficult pairs (*cabbage-pen, obey-inch, crush-dark, in-also*). In the standard procedure for scoring the test, the results for easy and hard items are summed. In the present study, they were treated as separate scores. For the results to be used in the analyses, the patient had to be able to repeat single words.

Previous studies of aphasics with this test are not known to the author. The differences between aphasic subgroups in semantic associations have been described by Howes. He found defects in what he called "group B" aphasics (Wenicke) as opposed to "group A" aphasics (Broca) (Howes & Geschwind, 1964).

In her large study of head injuries acquired in World War II, Newcombe (1969) included a test of association learning in which the subjects learned three unrelated pairs of items. She found no greater deficit in left- than in right-hemisphere injuries, but patients with parietal injury did poorly. Milner (1962) found deficit in patients with left temporal lesions for the paired-associate task from the Wechsler memory scales (the same as those used here), but only patients with temporal lesions (left or right) were tested.

5.5.1.4. Nonverbal Immediate Memory. The block-pointing-span test was similar to one used by Corsi (1972) for the study of memory functions. On a square board (20 × 20 cm), 12 blocks were mounted in a random arrangement. Their dimension was 2 × 2 × 2 cm (see Figure 5.1). On the side facing the examiner, the blocks were numbered 1 through 12.

The patient was instructed to point to the blocks shown by the examiner in the same order. The examiner pointed to the blocks one by one at the rate of one block per second. The number of blocks pointed to by the examiner was increased by one until the patient failed two consecutive trials. The procedure was repeated with instructions to point to the blocks in opposite sequence to that shown by the examiner. The score was the number of items in the longest series of blocks pointed to correctly forward and the number of items in the longest series of blocks pointed to correctly backward.

The original block-tapping test by Corsi (1972) has nine blocks. The normal span for a control group with a mean age of 28 years was given as 4.6 by Corsi. The same test has been used by other researchers

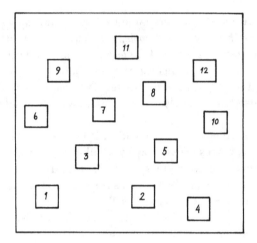

Figure 5.1. Block-pointing test.

who have mostly followed the procedure given by de Renzi and Nichelli (1975), in which two trials at every length are given and a half point is awarded for correct performance of the second trial. According to these authors, the presence or type of aphasia is not associated with a deficit on this test.

5.5.1.5. Nonverbal Serial Learning. After block-pointing span forward had been established, a new pointing sequence was constructed containing two items more than the span. The sequence was demonstrated repeatedly until a criterion of two consecutive perfect reproductions had been reached. Recall was attempted after each trial, and the number of correct items recalled was noted. The score was the number of trials to reach criterion.

In the patients scored for position effects on the verbal serial learning task, the same scoring was performed on the first two trials of the block-pointing sequence. The error types were evaluated in the same manner as with the digit learning task.

The same task (span + 2) was used by de Renzi, Faglioni, and Previdi (1977). This task was surprisingly difficult for their subjects, the controls needing 12.8 trials to reach the criterion. One suspects that their measure of span was the cause of the difficulty. Assume that a patient repeats all sequences of two, three, and four blocks correctly, but none of the longer sequences. The scoring system would give the patient a score of 6 points, and he or she would be required to learn a series of eight blocks. In the present study, the span would be scored as 4 points, and the learning task would be to repeat a sequence of six.

   5.5.1.6. Shape Association Learning. Six figures from the material of nonsense shapes constructed by Vanderplas and Garvin (1959) were selected. They were divided into three pairs. In Presentation 1, the pairs were presented to the patient simultaneously, each column forming a pair. The upper row was designated *stimuli* and the lower *responses.* The instructions were that the examiner had made an arbitrary decision that certain figures belonged together. The subject was to inspect the array for 30 seconds and then try to remember which shapes went together. The response shapes were then removed, and the stimulus shapes were rearranged in the left-to-right order (Presentation 2). The subject was handed a test shape (T) and was asked to match it to the corresponding stimulus. The response was placed with the selected stimulus, and a new response item was presented. When three choices had been made, the examiner rearranged the pairs into the correct combinations. This terminated the trial. The subject was allowed to examine the correct arrangement briefly before the next trial. The procedure was repeated until two consecutive trials had been performed correctly. The steps of the procedure are summarized in Figure 5.2. The score was the number of trials to criterion.

   The Vanderplas and Garvin stimulus material has been used in previous studies with aphasics, but not administered in the same way. Cermak and Tarlow (1978) used the material in a continuous recognition-memory paradigm and found it too difficult to be informative. De Renzi, Faglioni, and Villa (1977) used it in a study asking patients to sort eight patterns in a prescribed sequence. The left-hemisphere group, of which half were aphasic, performed only marginally worse than normal controls.

5.5.2. The Structure of Memory in Aphasia

   As noted, the relation of memory tasks to memory functions is complex, and the question of which memory functions are measured by the tests and the derived scores in the aphasic group is highly

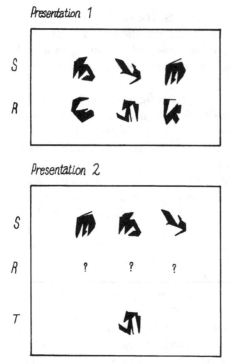

Figure 5.2. Shape association test.

pertinent. Therefore, the author, in collaboration with K. Sundet, performed factor analyses, first on the set of verbal and nonverbal measures separately and then on the combined set of measures. The method used was principal-component analysis to identify factors, an examination of eigenvalues or scree test (Cattell, 1978) to decide how many factors to include, and varimax rotation to determine factor loadings.

For the verbal scores, four factors with eigenvalues higher than 1.0 were found, and they accounted for 63% of the variance. The contributions of different tests to the factors are shown in Table 5.2. In general, a test is listed only on the factor to which it gives the highest contribution. When the values are close, the same test is listed under two factors.

The following interpretation may be suggested of the factors: Factor 1 is an immediate memory factor. In addition, the word-

Table 5.2. Factor Composition of Verbal Tests

Factor 1	Factor 2	Factor 3	Factor 4
DF (.81)	DSP-F (.62)	DS-T (.71)	DSP-M (.79)
DB (.78)	DSE-R (.84)	DSE-P (.64)	DSP-L (.39)
PS (.78)		DSE-S (.64)	DSE-I (.60)
WPA-A (.75)			
WPA-B (.56)			

association tests load on this factor. These tests were performed poorly, and the results may indicate that the main basis of performance was rote immediate memory.

Factor 2 is a serial learning factor with high loading on the error type of refusal and on performance on the first item. These features are shown elsewhere to be associated in nonfluent aphasics. Factor 3 is another serial learning factor with trials, perseveration, and sequence errors loading highly. The two latter features are shown elsewhere to be characteristic of fluent aphasia. Factor 4 is represented by serial learning measures of performance on the middle and last items and errors of intrusion. These features may be associated with a general failure to learn over trials and, more specifically, with a recency effect, although the latter is weak.

The factors account for all verbal measures, least satisfactorily for performance on the last item in digit serial learning, which loads only .39 on Factor 4.

The results for nonverbal tests are shown in Table 5.3. The criterion of eigenvalues above 1.0 results in five factors accounting for 68% of the variance.

The first factor is a nonverbal immediate memory factor with high loading on blocks forward and backward. Factor 2 is a serial

Table 5.3. Factor Composition of Nonverbal Tests

Factor 1	Factor 2	Factor 3	Factor 4	Factor 5
BF (.84)	BS-T (.67)	BSE-I (.58)	BSP-M (.78)	BSP-F (.86)
BB (.76)	BSE-P (.68)	BSE-R (.81)	BSP-L (.60)	SA-T (.43)
SA-T (.41)	BSE-S (.80)			

learning factor with trials associated with perseveration and sequential errors. It shows an interesting parallel to Factor 3 in the structure for verbal memory. Nonverbal Factor 3 associates two error types: intrusion and refusal. Factor 4 associates performance on the middle and last items on the block serial learning. Factor 5 is best represented by one single measure, the performance on the first item of the block serial learning. The factor solution accounts for all nonverbal measures, but most poorly for paired-shape association, which loads only .41 on Factor 1 and .43 on Factor 5.

In the design of the tasks, an attempt was made to construct parallel tests for measuring memory for verbal and nonverbal material. The results show that the intention was fulfilled by revealing a parallel structure of memory performances of aphasics.

The combined analysis of verbal and nonverbal tests yielded, in all, 10 factors with eigenvalues above 1.0. The scree test (Cattell, 1978) was used to limit the number of factors studied with varimax rotation, and five factors were included, accounting for 49% of the variance. Tests and factor loadings are shown in Table 5.4.

Factor 1 incorporates Factor 1 of the verbal factor analysis. In addition, intrusion errors on digit serial learning and paired-shape association are included. It is reasonable to maintain the interpretation of this factor as mainly a verbal immediate-memory factor. Factor 2 shows an interesting coupling of Factors 1 and 2 from the nonverbal analysis with Factor 3 from the verbal analysis. Immediate memory for block sequences shows the highest loadings, but the factor is also represented by other measures relating to learning and reproducing

Table 5.4. Factor Composition of Combined Verbal and Nonverbal Tests

Factor 1	Factor 2	Factor 3	Factor 4	Factor 5
DF (.58)	BF (.68)	DF (.58)	BS-T (.64)	BSP-M (.75)
DB (.54)	BB (.62)	DB (.46)	BSE-R (.54)	BSP-L (.51)
PS (.66)	BS-T (.54)	DSE-R (.77)	BSP-F (.38)	DSP-M (.34)
WPA-A (.78)	BSE-P (.59)	DSE-S (.40)		DSP-L (.38)
WPA-B (.66)	BSE-S (.61)	DSP-F (.61)		
DSE-I (.62)	DS-T (.47)			
SA-T (.53)	DSE-P (.40)			
	DSE-S (.40)			

sequences of verbal and nonverbal material. The finding of a sequencing factor is interesting in relation to the claim that sequencing is a distinctive error category in aphasia (Tzortzis & Albert, 1974). In studies of normal serial learning, it has been shown that a schema of the sequential structure of the list is learned independently of verbal associative relations (Ebenholz, 1972).

Factor 3 is a verbal serial learning factor similar to Factor 2 of the verbal analysis, except that some of the immediate memory measures appear again. Factors 4 and 5 are mainly nonverbal learning factors relating to error types and position effects. Under these factors, BSP-F, DSP-M, and DSP-L have been listed because this is where they show their highest factor loadings within the present solution.

It is to be expected that more of the specific factors of the verbal and nonverbal solutions will appear if a greater number of factors are analyzed. The present analysis is, however, sufficient to bring out the important point that material specificity is only partly preserved. Some factors are relatively purely material-specific, whereas some, notably Factor 2, combine measures by a different principle. The complexity is also brought out by the fact that tests may well show moderate factor loadings on several factors, some material-specific and some not. An example is DSE-S, which loads moderately on both Factor 2 and Factor 3 (a sequential factor and a verbal learning factor).

5.5.3. Relations of Memory to Aphasia Group

In these analyses, tests are grouped according to the functional domains suggested in Table 5.1, whereas the factor structure is taken into account in the discussion. In the analyses, the age bias has been corrected for statistically. The results are summarized in Table 5.5 and 5.6, and a brief discussion of each functional area follows. In Table 5.5, the nonverbal functional areas are not shown because no significant relationships were found.

5.5.3.1. Verbal Immediate Memory. On all types of tests, there were significant main effects of type and severity of aphasia (Tables 5.5 and 5.6). In general, nonfluent aphasics performed worse than fluents on all tests. Significant interactions were found only on DF. The interaction of type and severity ($F = 2.81$, $p < .05$) was caused

Table 5.5. Summary Table of Tests Showing Relation to Type of Aphasia

	Test with relative deficit	
Function	Nonfluent	Fluent
Verbal immediate memory	DF ($p < .01$) DB ($p < .05$) PS ($p < .05$)	
Verbal serial learning	DSP-F ($p < .05$) DSE-R ($p < .01$)	DSE-P ($p < .05$) DSE-S ($p < .05$)

by the relatively poor performance of mild nonfluents as compared with mild fluents. The observed interaction confirms the finding of Goodglass, Gleason, and Hyde (1970) that Broca aphasics have strikingly poor immediate memory span, except that their study used a test closely similar to PS. The other significant interaction was severity and chronicity ($F = 7.18$, $p < .01$). The main source of the interaction appeared to be that the mildly aphasic acute cases performed disproportionally better than the mildly aphasic chronic cases.

Table 5.6. Summary Table of Tests Showing Relation to Severity of Aphasia

Function	Tests with deficit in severe aphasia	
Verbal immediate memory	DF	($p < .001$)
	DB	($p < .001$)
	PS	($p < .001$)
Verbal serial learning	DSP-F	($p < .01$)
	DSE-I	($p < .01$)
	DSE-R	($p < .01$)
Verbal associative learning	WPA-A	($p < .001$)
	WPA-B	($p < .001$)
Nonverbal immediate memory	BF	($p < .03$)
	BB	($p < .001$)
Nonverbal serial learning	BS-T	($p < .01$)
Nonverbal associative learning	SA-T	($p < .01$)

5.5.3.2. *Verbal Serial Learning.* The number of trials to criterion yielded no significant main effects or interactions. Interesting group differences appeared, however, both in error types and in position curves for the first two trials. The position curve data show a superiority of fluent over nonfluent aphasics in recalling the first item, but not the middle or the last.

No significant main effect or interaction was shown on the final item, but the visual form of the curves gives the impression of an absence of recency effect in the fluent group, especially the severe fluents (Figure 5.3). A direct test of the significance of the difference between the first and the middle and between the final and the middle positions was therefore performed within all groups (Table 5.7). An analysis of variance was performed to test for an overall effect of serial position on memory score. If the resulting F score was statistically significant ($p < .05$), then the differences between the initial versus the middle and the last versus the middle positions were tested for significance in order to reveal the specific locus of the position effect. The results show that all groups had a significant primacy effect, but only the nonfluent groups had a significant recency effect.

Despite an apparent advantage on the first two trials in coding information in long-term storage, the fluent patients performed no better than the nonfluents. The explanation may be their tendency to make errors of perseveration and sequential ordering, errors related exclusively to type and not to the severity of the aphasia.

On intrusion errors, in addition to the significant main effect of severity there was an interaction of type and severity. The difference between mild and severe was particularly striking for fluent patients ($F = 2.66, p < .05$).

Refusal to continue was most characteristic of nonfluent patients. In addition to being significantly more common than in fluent patients, it was also the most frequent error type in nonfluents. The result is consistent with the finding that a so-called catastrophic reaction (depression, rejection, withdrawal) is a reaction type found more frequently in this group (Robinson & Benson, 1981). This type of reaction is, however, also strongly related to severity of aphasia, and the two effects are independent.

5.5.3.3. *Verbal Associative Learning.* The results (Table 5.7) show a strong relation to the severity of the aphasia and no significant interactions.

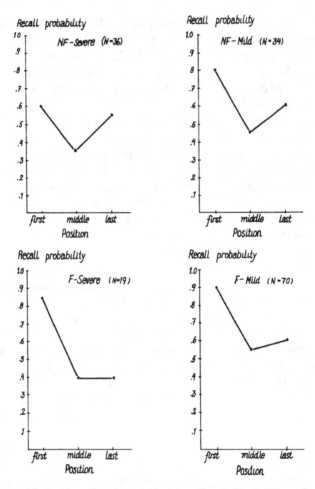

Figure 5.3. Position effects in serial learning. Key: F, fluent; NF, nonfluent.

5.5.3.4. *Nonverbal Immediate Memory.* The results show a strong relation to severity but no significant relation to or interaction with other dimensions.

5.5.3.5. *Nonverbal Learning.* The score on trials on block serial learning was related to the severity of the aphasia. There were no significant differences between groups in performance on the first,

Table 5.7. Within-Group Position Effects

	Digits				Blocks			
	Position effect				Position effect			
	F	p	I>M	L>M	F	p	I>M	L>M
Severe nonfluent	3.29	.04	.01	.02	4.50	.01	.02	n.s.
Mild nonfluent	7.03	.01	.001	.05	2.36	n.s.	—	—
Severe fluent	5.86	.01	.01	n.s.	1.90	n.s.	—	—
Mild fluent	22.18	.001	.001	n.s.	5.42	.01	.01	n.s.

Note. I = initial item; M = middle item; L = last item.

the middle or the last items. As shown in Table 5.7, two of the groups (severe nonfluent and mild fluent) showed a significant position effect, and the other two did not. In both cases, the significant effect was a primacy effect.

The error types on block serial learning showed no significant relation to either the type, the severity, or the chronicity of the aphasia. The lack of a significant relation does not mean that errors were not made. The data show sequential error to have been present in 40% of the cases, perseveration in 25%, intrusion in 26%, and refusal in 5%.

The results on paired-shape association show significant relation to severity but not to other dimensions of aphasia.

5.5.4. Discussion

For a task to show a material-specific deficit, it should show no difference from a control group when the noncritical material is tested, and the performance on noncritical material should be uncorrelated with performance on critical material.

In the study by de Renzi and Nichelli (1975) of the block-pointing test, both conditions were satisfied. In the present study, only the latter hypothesis was tested, and the condition for material specificity was not satisfied. Both the block-pointing test and all other nonverbal tests showed a highly significant relationship to severity of aphasia. The explanation of the discrepancy is probably connected with the

likelihood of detecting weak statistical effects. The highly significant
F ratios relating severity of aphasia to nonverbal memory and learning
defects correspond to correlations between .27 and .36, and the high
number of subjects tested, including many severe cases, contributed
to the significant result.

The complexity of the material-specificity issue is further clarified
by the factor analyses, which show that no extreme position can be
taken for or against specificity.

When subgroups are examined the material specific nature of
the deficit is more sharply focused. Whereas the performance with
verbal material shows several features related distinctly to type of
aphasia, this is not true of nonverbal material. In no case is the nonverbal deficit related to type of aphasia. In the direct test of relations
to groups, there is no indication that specific error types in serial
learning or differences in the initial, the middle, and the last position
on the serial position curve found for verbal material are also found
for nonverbal material. The factor analyses, however, revealed parallel features of organization across material types.

With verbal material, the pervasive effect of severity of aphasia
is striking on all tests except digit serial learning, both trials and error
types. This exception may be taken to indicate that the procedure of
adjusting the length of the digit series to the memory span rather
than using a series of fixed length for all subjects is successful in
eliminating nonspecific severity effects. When adjusted in this way,
the tasks yield valuable differential information.

The evidence for subtypes of deficits in verbal memory comes,
first, from tasks of immediate memory, which were performed more
poorly by nonfluent than by fluent aphasics. The fact that mild nonfluents had a strikingly poor performance was noted above. From an
examination of the serial position curves, it appears further that nonfluents were inferior to fluents on the first item of a digit series, but
not on the middle or final items. The error type most characteristic
of nonfluent patients was refusal to continue.

Fluent aphasics had a better immediate memory span than nonfluents and better retention of the initial item of a series of digits.
Still, they failed to show an overall advantage in verbal learning. In
the serial position curve, they failed to show an advantage of the final
compared to the middle item in recall, thus supporting the idea that

they had a weak acoustic memory (precategorical acoustic storage). They had a tendency to make certain error types, specifically perseveration and sequential errors. Assuming that fluent aphasics have posterior lesions, one cannot take the result to confirm the notion frequently expressed that perseveration is a sign of frontal lobe pathology (Luria, 1966). The findings suggest that it is not so much perception of sequence but response organization and reorganization that fails in the fluent aphasics. The nonavailability of the final item as an anchoring point for response organization may be critical. This hypothesis is supported by the fact that the relation of these error types to fluent aphasia disappeared in the block-pointing serial learning task, in which there was also no difference between aphasia types in performance on different serial positions.

The verbal association tests yielded no differential information, and the expected semantic impairment in fluent aphasics could not be demonstrated. Chronicity did not come out as a significant main effect in any analysis. There is, thus, no type of memory deficit that is more characteristic of acute aphasics than of chronic.

A search of the aphasia registry for patients with mild aphasia, low verbal immediate memory, and high nonverbal immediate memory yielded no cases among the 249 patients analogous to those reported by Warrington and Shallice (1969) or by Basso *et al.* (1982).

5.5.5. Conclusion

I suggest in conclusion that severe aphasia is associated with reduced encoding of information. This is more true of verbal than of nonverbal material, but only relatively so. The view that verbal memory problems are just secondary to a general verbal encoding deficit (severity of aphasia) must be rejected because there are clear indications of subtypes, as described above.

The indications of specificity found in the material show that the view implied in the material-specificity thinking—namely, that tasks can be decomposed into independent components—does not hold. The factor analyses give the impression of multiple, overlapping organization of function, so that a given aspect of a task may be represented both in a purely verbal factor and in factors more related to the structural properties of tasks (e.g., sequential organizations).

When the severity dimension of aphasia is analyzed, the results seem to reflect the deficit in multiple dimensions of organization of memory. The memory deficits in severely aphasic patients are thus probably complex, reflecting a verbal deficit, a deficit of spatiotemporal encoding, and a deficit of response organization.

DEFECTS OF VISUAL
NONVERBAL ABILITIES

The functional domain covered by the above title is not well defined. Excluding tests of memory it is not clear how theories of normal mental abilities would structure the domain of remaining nonverbal functions.

Taking a logical approach, we may classify tasks as varying in three dimensions, complexity of stimuli (spatial configuration), complexity of response (coordinated movement), and complexity of intervening functions (logical principle for linking stimulus and response). We may then hypothesize that deficits will reflect this dimensional structure, and this hypothesis can serve to structure the discussion of deficits in aphasia.

6.1. Visual Nonverbal Functions in Aphasia

It is doubtful if perceptual complexity in itself accounts for a dimension of the deficit in aphasia. Clinical studies have found deficits in tests combining perceptual complexity and either manipulative responses or symbolic complexity. In the first case, the deficits are classified as visuoconstructive and measured by performances in drawing or puzzle-type tasks. There is an extensive literature on the performance of aphasics on such tasks (see de Renzi, 1982, Chapter 9). Although it is acknowledged that patients with right-hemisphere injury perform such tasks poorly, the discussion has been about whether there is a difference in performance in favor of left-hemisphere-injured patients. Some studies (Arena & Gainotti, 1978) indicate that this is not the case. The question of qualitative differences between left- and right-hemisphere injuries is complex. Hécaen and Assal (1970) hypothesized that the deficit in left-hemisphere injuries was on the executive rather than the perceptual side and found that guidemarks aided the copying of a cube in left- but not in right-hemisphere injuries. Other qualitative differences noted in the literature are lack of detail with preserved spatial organization in left-hemisphere injuries and distorted spatial organization and neglect of the left side in right-hemisphere injuries. The studies summarized by de Renzi (1982) were unable to consistently quantify or reproduce these clinical observations.

Moreover, the idea that the deficit is executive and not perceptual is contradicted by the failure to reproduce the findings of Hécaen and Assal (1970), as well as by low scores on spatial perceptual tests with multiple-choice alternatives in both left- and right-hemisphere-injured patients with constructional apraxia (Arena & Gainotti, 1978). Further evidence for a deficit of perceptual function comes from the studies of the Gottschaldt hidden figures test (Teuber & Weinstein, 1956; Russo & Vignolo, 1967), which is performed more poorly by aphasics than by any other group with localized injuries. The test undoubtedly also makes intellectual demands and was interpreted as showing an intellectual deficit in aphasics by Teuber and Weinstein (1956). Tests suitable for demonstrating an intellectual deficit should not require elaborate instruction, the stimuli should not be complex, and the responses should be simple. Weinstein (1964) summarized a series of studies showing nonverbal deficits, including visual conditional reaction, a paradigm taken from animal learning. Carson *et al.* (1968) found a normal acquisition curve for simple identification learning, but inferior performance on an alternative reaction task. These findings are consistent with the observation that even severe aphasics (globals) can learn to attach meaning to stimulus cards with abstract shapes (Gardner, Zurif, Berry, & Baker, 1976). The failure to perform more complex tasks (conditional or alternating reaction) may indicate an intellectual disturbance. Further evidence comes from a study by

Basso, de Renzi, Faglioni, Scotti, and Spinnler (1973), who, after controlling for confounding influences with a covariance technique, concluded that there is a specific deficit of visual-intellectual abilities with posterior left injuries. A newer study by Basso *et al.* (1981) did not clarify the issue further. Other studies with the Raven test (Kertesz & McCabe, 1975) found a moderate correlation between test score and degree of aphasia.

Studies with the Wechsler intelligence scales (WAIS; Wechsler, 1958) reported poor performance in aphasics (Orgass, Hartje, Kerschensteiner, & Poeck, 1972). For some scales, the results were worse for aphasics than for right-hemisphere-injured patients. The problem with interpreting these results is that many of the patients performing poorly showed constructional apraxia. Indeed, some of the subtests of the WAIS (Block Design) have been used as instruments for measuring constructional apraxia (e.g., Black & Strub, 1976). Some authors have therefore corrected for constructional apraxia before assessing the relationship of aphasia to "intelligence" (Borod, Carper, & Goodglass, 1982). If this correction is made, the effect of aphasia tends to vanish. In this context, I have tried to avoid a discussion of "intelligence" and have used the more neutral term *visual nonverbal abilities*. I agree with the position of Hamsher (1982), who stated "it is not clear how one can separate constructional praxis from nonverbal intelligence" (p. 344). The point may be not to control for constructional praxis, but to point out that aphasics show neither a general perceptual nor a general intellectual deficit. A more specific association of these factors is required to bring out a deficit.

6.1.1. Apraxia

Although the visual-intellectual deficits are multidimensional and difficult to disentangle, there is good evidence of a specific deficit in the execution of motor activities with left-hemisphere injuries. The term *apraxia* was coined by Liepmann (1900), who distinguished between ideational and ideomotor apraxia. The concepts are intimately connected with a hierachical concept of control centers in execution of motor acts. The disconnection of language areas from

motor control pathways can account for some cases of apraxia (Geschwind, 1967a), but in addition, it is necessary to infer higher order
mechanisms for the control of skilled acts in the left hemisphere
(Kimura, 1979). Aphasia and apraxia are frequently associated (Kertesz & Hooper, 1982; Poeck & Lehmkuhl, 1980).

There are two forms of explanation for the association of aphasia
and apraxia, and for the association of aphasia with a nonverbal deficit
in general. The first is the assumption of a common-core deficit underlying both verbal and nonverbal deficits. Kimura (1979), following
Liepmann (1900), argued for this conclusion in discussing the relation
between aphasia and motor disorders.

To the apparent counterargument that not all apraxics are aphasics and vice versa, the reply would be that clinical tests are not
sufficiently sensitive to reveal mild deficits. A more telling argument
has been given by Poeck and Huber (1977), who said that the linguistic
aspects of aphasia are left unexplained by the hypothesis.

Another version of the common-core hypothesis sees aphasia
as a defect of symbolic activity, asymbolia (Finkelnburg, translated
by Duffy & Liles, 1979). Of particular interest is the defective use and
understanding of gesture and pantomime in aphasics (Duffy & Duffy,
1981; Varney, 1978). Cicone, Wapner, Foldi, Zurif, and Gardner (1979)
found a similarity in the qualitative aspects of gesturing and speaking.
This was not found by Lehmkuhl, Poeck, and Willmes (1983) when
they analyzed error types in tests of apraxia.

Opponents of the common-core hypothesis point out that it
makes the strong assertion that phenomena are invariably associated,
and that if exceptions exist, then the hypothesis can be rejected. They
further point out that, if the phenomena are both present but are not
correlated (Goodglass & Kaplan, 1963) or show different recovery
rates (Poeck & Lehmkuhl, 1980), then they are not manifestations of
a single function. Both Goodglass and Kaplan (1963) and Poeck and
Lehmkuhl (1980) advocated the so-called anatomical hypothesis, saying that the proximity of the neural substrate makes for clinical association because of the typically large lesions occurring in common
forms of pathology. The second hypothesis is the more conservative
and should be adopted. Even so, it is legitimate to speculate that
the proximity of brain representation may give a clue to functional
relatedness or evolutionary association. Although assertions about

evolution are necessarily speculative, it seems likely that verbal (auditory-vocal) language evolved in a context in which visual-gestural communicative and symbolic abilities already existed.

6.2. The Present Study

A number of tests representative of traditional operationalizations of constructional deficits, apraxia, and nonverbal reasoning were performed. It was hypothesized that this three-dimensional structure is reflected in an aphasic population. The hypothesis was tested with factor analysis and an ensuing analysis of variance relating the test results to type, severity, and chronicity of aphasia. Other researchers have reached diverging conclusions on the relation of nonverbal deficits to aphasia classification. Some have found no relation to severity and some a strong association (see above). If the type of aphasia is considered, the authors finding a relationship to type generally ascribe the most severe deficits to patients with posterior lesions and fluent speech, but the results have not been consistent. The attempt to avoid confounding the type and the severity of the aphasia, which is an important feature of the present system, may clarify the relationships.

6.2.1. Tests of Nonverbal Abilities

6.2.1.1. *Raven Coloured Progressive Matrices (Raven, 1960).* This test (RCPM) consists of a test booklet with 36 tasks. Each task contains a pattern with an empty slot and six alternative patterns that all fit into the slot, but with different designs. For each task, only one of the alternatives is a correct response, and the subject is asked to choose. The tasks are divided into three series of increasing difficulty (A, Ab, and B), starting with perceptual matching problems and progressing toward more abstract reasoning problems. In some studies, it has been useful to distinguish between the series as measures of different underlying functions (Denes, Semenza, Stoppa, & Gradenigo, 1978; Costa, 1976).

The test is likely to show a greater deficit in patients with posterior than in patients with anterior lesions (Basso *et al.*, 1973). It does not discriminate well between patients with right- and left-hemi-sphere lesions (Arrigoni & de Renzi, 1964). A moderate but significant correlation with aphasia was found by Kertesz and McCabe (1975), but not by de Renzi and Faglioni (1965). All in all, it is, however, generally regarded as a test that may often be performed very well even by patients with a severe aphasia (Zangwill, 1964). Among aphasic subgroups, the ones with the most severe aphasia also show the highest incidence of subnormal RCPM scores (Kertesz & McCabe, 1975).

  6.2.1.2. Wechsler Adult Intelligence Scale, Performance Test. This test (WAIS) is a commonly used psychometric test for adults, and it has also found application in clinical neuropsychology (McFie, 1975). The performance tests consist of five scales.

  Digit Symbol (DS) involves filling empty boxes with written sym-bols according to a digit–symbol code. The test demands writing with the left hand in aphasics with right-side hemiparesis and therefore introduces undesirable possibilities of contaminating factors. In the present study, this test was omitted and each patient was assigned a score corresponding to the average of the other performance scales.

  Picture Completion (PC) consists of 20 pictures, each with a miss-ing detail. The missing detail must be named or pointed out. Accord-ing to McFie (1975), this is the performance test showing least alterations after cerebral injury.

  Block Design (BD) uses nine blocks with red, white, and half-white and half-red sides. Patterns must be constructed by joining the blocks. Nine patterns of increasing difficulty must be reproduced. According to McFie (1975), this test is maximally sensitive to parieto-occipital lesions of the right hemisphere.

  Picture Arrangement (PA) consists of cartoonlike picture series where the subject must arrange the pictures in proper sequence. There are 12 series. This test is maximally sensitive to frontotemporal injury of the right hemisphere, again according to McFie (1975).

  Object Assembly (OA) consists of five puzzles of naturalistic designs. The results are often parallel to those for Block Design.

  There are many studies showing a pattern of verbal tests' being performed more poorly than performance tests in left-hemisphere injuries, and the reverse pattern in right-hemisphere injuries. From this pattern, it does not follow in general that patients with left-hemisphere injuries perform normally on performance scales. Orgass *et al.* (1972) showed that, on the average, aphasics perform as poorly on performance tests as patients with right-hemisphere injury, whereas patients with left-hemisphere injury without aphasia perform better than both other groups. There are differences in pattern of performance on subtests, however, aphasics performing most poorly in relation to other groups on picture completion.

6.2.2. Tests of Motor Function

Finger Tapping (FT-L) is from the Halstead–Reitan neuropsychological battery (Reitan & Davison, 1974) and is a telegraph-type key connected to a counter. The subject rests his or her hand on the table and with the index finger depresses the key as many times as he or she can in 10 seconds. The average of five trials is recorded. Only the scores for the left (unimpaired) hand are used.

Grooved Peg-Board (GP-L) is a test from the Halstead–Reitan neuropsychological battery. It consists of a board covered with a metal plate containing a 5 × 5 matrix of holes. An adjoining cup contains 30 pegs. Each hole has a groove oriented in different directions, and each peg has a tag. The peg must be appropriately oriented with respect to the grooved holes in order to be inserted. The time taken to fill all the holes is recorded. If the subject looses a peg, an error is counted. In this study, only results with the hand ipsilateral to the injured hemisphere were used in the analyses.

6.2.3. Apraxia

The set of disturbances known as *apraxia* is believed to encompass several subtypes. The tests employed vary accordingly. Tests of constructional ability (constructional apraxia) are

> *Copy-a-cross (COPY)* from the Halstead–Reitan battery, range 0 to 5 (Reitan & Davison, 1974).
> *Frostig Copying (FROS)* from Frostig (1966), range 0 to 100.

Tests of imitative movements (ideomotor apraxia) were designed with reference to Luria (see Christensen, 1975) and Goodglass and Kaplan 1963, 1972):

Imitative Finger Position (FING) with the left hand, range 0 to 20. *Hand Movements Imitation (MOV-I)* with the left hand, range 0 to 34.

Tests of ideational apraxia consist of

Manipulating Real Objects (OBJ), range 0 to 16.
Picture Sequencing (PICT), in which photographs of daily activities must be ordered correctly, range 0 to 17.

6.2.4. Results

A factor analysis with principal component analysis and varimax rotation was performed to clarify the structure of performances. The first factor accounts for 49% of the variance in the unrotated solution. The second, but not the third, factor has an eigenvalue above 1.0. Because a three-factor solution was looked for, the loadings of the different tests on the three first factors after rotation are still shown in Table 6.1. The factors account for 64% of the variance.

All the nonverbal intelligence tests come out as the first factor. Apraxia can be clearly distinguished from these in the second factor, and Factor 3 reflects simple speed and copying performances. Solutions for four and five factors were calculated to see which of the three factors would remain intact. The first two factors are unaffected by the more comprehensive solutions.

Table 6.1. Three-Factor Solution of Performance Structure

Factor 1	Factor 2	Factor 3
RAV-A (.76)	FING (.71)	FT-L (.71)
RAV-Ab (.84)	MOV-I (.77)	
RAV-B (.86)	OBJ (.88)	FROS (.46)
PC (.53)	PICT (.64)	COPY (.61)
BD (.68)	GP-L (.59)	
PA (.59)		
OA (.66)		

Table 6.2. Relation of Visual Nonverbal Ability to Severity of Aphasia

Function	Test related to severity
Nonverbal intelligence	RCPM ($p < .001$) Wechsler PIQ ($p < .001$)
Apraxia	GP-L ($p < .01$) MOV-I ($p < .001$) FING ($p < .001$) OBJ ($p < .001$) PICT ($p < .01$)
Others	Copy ($p < .01$)

6.2.4.1. Relation to Aphasia Classification. Because all the sub-scales of the Raven and Wechsler (PIQ) load highly on the first factor, it is reasonable to analyze the sum scores only in relation to the aphasia group. In the following tables, the results have been corrected for the correlation of aphasia severity with age. The results for the functional areas suggested by the factor analyses are summarized in Tables 6.2 and 6.3. For nonverbal intelligence tests, the relationship to severity of aphasia is significant and corresponds to a correlation of .38 and .27 for the Raven and Wechsler measures, respectively. A tendency of the fluent aphasics to score somewhat better on the Raven test than the nonfluent is not significant. There are no significant inter-actions between the classification variables.

For the tests of apraxia, the relationship to severity of aphasia was consistently present, corresponding to correlations from .27 (PG-L) to .51 (FING). For tests of imitating hand or finger movements,

Table 6.3. Relation of Visual Nonverbal Ability to Type of Aphasia

Function	Test with deficit	
	Nonfluents	Fluents
Apraxia	MOV-I ($p < .01$)	

there was an association with type of aphasia, and the tendency was for nonfluent patients to perform worse than fluent patients. These are the tests most closely reflecting the concept of ideomotor apraxia. The tests measuring ideational apraxia (OBJ and PICT) showed no relation to type of aphasia. There were no interactions of the classification variables with respect to any of these tests.

Factor 3 was taken to reflect severe constructional difficulty coupled with slowness of repetitive finger movements.

For tests encompassed by this factor, the relationship to any dimension of classifying aphasia was weak or absent.

6.2.5. Discussion

The three factors isolated are interpretable in relation to previous discussions of nonverbal deficits as measuring nonverbal ability, apraxia, and constructive deficits. The latter can be defined separately from the ability factor when gross deficits on simple copying tasks are used as a criterion, but within the ability factor, a split between logical reasoning and more executive aspects of ability was not found, even when four and five factor solutions were analyzed.

For the tests in the first factor, a correlation with severity of aphasia of the same magnitude as previously reported by Kertesz and McCabe (1975) was found. From the study of Basso et al. (1973), one might have expected more severe deficits in fluent than in nonfluent patients, but these were not found.

The tests of apraxia showed a clear correlation with severity of aphasia, confirming the results of Kertesz and Hooper (1982) and contradicting those of Goodglass and Kaplan (1963) and of Lehmkuhl et al. (1983). Imitation tasks of moderate complexity (finger and hand positions and intransitive movements) were especially difficult for nonfluent aphasics. This type of task was not tested by Kertesz and Hooper (1982). Lehmkuhl et al. (1983) found no differences between aphasia types.

The disturbances of simple executive left-hand functions were not related to the aphasic disturbance. They may have reflected functioning of the right hemisphere. It was impossible without routine CT-scans, to exclude patients who may have had right hemisphere lesions, and they may have accounted for the findings. So far, the results are consistent with what has been termed the *anatomical hypothesis*, that language areas and areas underlying nonverbal functions have proximal localization. I shall, however, discuss the issue again when the recovery data have been presented, considering the possibility that neither the common-core nor the anatomical hypothesis is correct. What we are observing may be a multidimensional response of the preserved brain to an injury in which a coupling of nonverbal and verbal factors is apparent in severe aphasia. This coupling may reflect the functional state of the preserved brain.

LOCALIZATION OF
LESION IN APHASIA

7.1. Status of the Localization Model

Our current aphasia classifications have developed in a context of
neurological diagnosis. The possibility of drawing firm conclusions
on the localization and type of pathology was a central concern in a
classification system. With the introduction of refined neuroradio-
logical methods for localizing lesions, there has been renewed interest
in localization studies. The major concern of such studies today is to
answer whether accepted views on the neurological basis of language
functions are essentialy correct.

In this context three possible conclusions may be envisaged:

1. The Wernicke–Lichtheim model is correct and sufficient to
explain the phenomena of aphasia. The specific assumptions of the
model were stated in Chapter 2.

2. The Wernicke–Lichtheim model is incomplete. There are
additional areas that deserve to be called language areas in the sense
that lesions of these areas give rise to aphasia. The resulting types of
aphasia are different from the classical syndromes.

3. The Wernicke–Lichtheim model is correct in stating that the
likelihood of aphasia is high with lesions of the classical language
areas. It is wrong in ascribing the variations in aphasia types to lesions
of differently localized modules within the language areas. It is the
pattern and volume of lesions combining areas within the classical

language areas with neighboring areas that determine symptom formation. The latter areas thus interact intimately with the language areas but do not, in most cases, give rise to aphasia when lesioned in isolation.

Briefly summarized, the first conclusion says that all is well with classical aphasiology. Position 2 says that the conceptual schema of the classical model is valid and can be extended to new areas and new forms of aphasia. Position 3 says that localization must, to some degree, be supplemented by a concept of interactive processing.

Early studies with CT-scan or isotope localization stressed the essential correctness of the Wernicke–Lichtheim model (Naeser & Hayward, 1978; Kertesz, Lesk, & McCabe, 1977; Kertesz, Harlock, & Coates, 1979). During the last few years, reports on aphasia with lesions outside the classical language areas have appeared with discussions of the possible mechanisms (see below).

7.2. New Candidates for Status as Language Areas

7.2.1. The "Limbic System"

The term *limbic system* is used to include the limbic lobe and the associated subcortical nuclei. The limbic lobe includes the subcallosal, cingulate, and parahippocampal gyri, as well as the underlying hippocampal formation and dentate gyrus. The main subcortical nuclei associated with the limbic lobe are the septal nuclei, the amygdaloid complex, the hypothalamus, the epithalamus, and various thalamic nuclei.

Indications that the limbic system has a role in language functions were reviewed by Lamendella (1977). The cingulate gyrus may be important in disturbances of language activation, with lesions sometimes giving rise to mutism (Robinson, 1976). The hippocampal area is considered of some importance in human recent memory. The work of Corsi (1972) indicates that lesions of the left hippocampus cause selective disturbances of verbal memory function. The possibility may therefore be considered that some of the variability in verbal memory in aphasics is caused by varying involvement or disconnection of the hippocampal region. There are no indications that the subcortical nuclei and fiber tracts of the limbic system are important to language.

7.2.2. The "Lenticular Zone"

The lenticular zone was defined by P. Marie (1906) and comprises a quadrangle of tissue extending from the anterior and posterior borders of the insular cortex to the midline of the brain.

Proceeding in the lateral-to-medial direction, the lenticular zone contains the following structures:

7.2.2.1. *Insula*. The insular cortex lies in the depth of the Sylvian fissure and is a conventionally defined neuroanatomical structure. It is roughly triangularly shaped and its outer limits are defined by the sulcus circularis. Wernicke (1874) attributed great importance to the insula as an association area for fibers from the frontal and temporal cortex. The view that lesions of the insula cause aphasia was, however, contradicted by Henschen (1922) on the basis of his review of 1,200 published autopsy reports. Penfield and Roberts (1959), as well as Rasmussen and Milner (1975), confirmed that no interference with language functions is found with electrical stimulation of the insula. The possibility remains that the insula interacts with other structures. Mohr (1976) stressed the importance of insular involvement in conjunction with lesions of the Broca area in producing the symptom complex of Broca aphasia.

7.2.2.2. *Capsula Extrema*. A fiber bundle believed to carry frontal-insular-temporal association fibers.

7.2.2.3. *Claustrum*. A sheet of gray matter. It has recently been shown to have extensive reciprocal connections with sensory cortical areas and particularly with visual areas.

7.2.2.4. *Capsula Externa*. A fiber bundle carrying mainly association fibers.

7.2.2.5. *Basal Ganglia*. The nucleus lentiformis is a prominent structure located laterally to the internal capsule. Medially, the head of the caudate nucleus can be seen bordering on the frontal horn of the ventricles. Traditional interpretations have stressed the primary motor functions of the basal ganglia, but Teuber (1976) urged a wider interpretation of their function. A role in the language function was considered unlikely by classical authors (e.g., Liepmann, 1915).

In recent studies, Damasio, Damasio, Rizzo, Varney, and Gersh (1982) and Wallesch, Kornhuber, Brunner, Kunz, Hollerbach, and Sugar (1983), among others, have reported on aphasia with basal ganglia lesions. Some of these cases have had lasting deficits.

7.2.2.6. *Internal Capsule*. A massive and compact layer of white matter carrying all fiber projections—afferent and efferent—between the cerebral cortex and the subcortical structures. These include fibers linking cortical areas by a cortex–basal-ganglia–thalamus–cortex loop.

7.2.2.7. *Thalamus*. Medial to the internal capsule lies the thalamus, which borders on the third ventricle. The pulvinar is a large thalamic nucleus located at the posterior end of the thalamus bordering on the posterior horn of the lateral ventricle. The pulvinar is of great interest in view of its massive connection with the posterior language area. The functional significance of these connections are, however, uncertain. Electrophysiological studies have shown interference with language functions after stimulation of the left, but not of the right, pulvinar (Fedio & van Buren, 1975). The stimulation effects are not generally different from those seen after cortical stimulation. The parallel results of stimulation studies may be taken to support a conception of the posterior language area and the pulvinar as one functional unit. It is possible that the stimulation effects are caused by indirect interference with cortical functions. However, a well-documented series of cases with subcortical lesions centering on the basal ganglia or the thalamus was presented by Alexander and LoVerme (1980). Penfield and Roberts (1959) interpreted the pulvinar as an important relay station between the posterior and the anterior language cortex. They assumed that the pathway is from the pulvinar to the dorsomedial nucleus of the thalamus and from that to the Broca area. Cases of thalamic aphasia have been published by, among others, McFarling, Rothi, and Heilman (1982). Luria (1977) suggested that the thalamus has important functions in connections with attentional control of the cortical language mechanism.

7.2.3. Medial Structures

The precise localization of the *supplementary motor cortex* in humans is not known. Electrophysiological work points to several discretely located sensorimotor representations of the body surface in animals

(Woolsey, 1958). In humans, a supplementary motor area was described by Penfield and Roberts (1959) on the basis of electrical stimulation. It is located just anterior to the motor strip, occupying part of the medial surface and extending onto the convexity of the frontal lobe. The limitations of the area rest on electrophysiological criteria, and anatomical extensions cannot be precisely defined.

The area is not considered in the classical literature on aphasia, not even in Henschen's review (1922) of all published cases with autopsies. The first hint of functional significance to language comes from the work of Penfield and Roberts (1959), who found interference with language after electrical stimulation in local anesthesia. Since then, some cases of aphasia with infarction of the supply area of the anterior cerebral artery have been published (Rubens, 1976). For review, see Razy, Janotta, and Lehner (1979) and Alexander and Schmitt (1980). Work with regional cerebral blood flow (Lassen, Ingvar, & Skinhøj, 1978) has also attributed a significant function in speech to the supplementary motor area. After lesions in this area, the characteristic defects most frequently pointed out are preserved repetition with a tendency toward echolalia, but paucity of spontaneous speech.

7.2.3.1. *Corpus Callosum*. This is a massive bundle of commisural fibers connecting the two hemispheres.

Lesions of the corpus callosum or of callosal fibers may produce disconnection syndromes in the form of inability to perform verbal commands with the left hand (a form of apraxia; Geschwind, 1967a) or inability to read material in the left visual field (a form of alexia; Benson & Geschwind, 1969). Modality-specific naming defects may also occur with callosal lesions. Hemispheric integration may be important in memory functions and in certain motor skills.

7.3. The Present Study

Analyses of a subsample of the present group of patients have been presented by Reinvang and Dugstad (1981) and by Reinvang (1983). In those studies, the relationship of aphasia types to localization was described in terms of the conditional probabilities of the results on one variable, given information on the other. Statistical analyses of differences in aphasia test parameters as a function of lesion localization were also reported.

Most analyses of lesion localization in aphasia have limited themselves to descriptive accounts in the form of case descriptions or composite lesion diagrams (Kertesz, Harlock, & Coates, 1979). Apart from the conditional probabilities used by Reinvang and Dugstad (1981), Blunk, De Bleser, Willmes, and Zeumer (1981) have presented a quantitative approach to analysis based on a subtractive method of lesion comparison in major syndromes. The present study uses a more refined quantitative method, that of canonical discriminant function analysis, in order to relate classification and lesion, and considers the merits of alternative classifications.

The further question asked is about the pathological basis of the performance on different parameters of the aphasia test. Classification systems rest on the assumption that the underlying parameters of classification are related to lesion localization.

Finally, the difficult question of the relative independence of a lesion effect is asked. Few of our patients had lesions restricted to only one anatomical region. The question is whether lesions of a given area have independent effects on performance irrespective of the context of other lesions. The Wernicke–Lichtheim model asserts that this is the case. An example is the explanation of global aphasia, which is seen as a mixed syndrome. The deficits characteristic of Broca and Wernicke aphasia (anterior and posterior lesions) are added together to obtain the mixed syndrome.

The example also serves as a starting point for criticizing the assumption of independent effects. Global aphasia has distinctive features not found in either Broca or Wernicke aphasia, notably stereotypy or recurring utterances. Fluency is generally poorer in a mixed anterior-posterior lesion than in a pure anterior lesion, although a posterior lesion in itself does not reduce fluency (Reinvang, 1983).

The analysis of interactions between lesions in the present study used the approach of comparing correlations and performances in groups with relatively discrete or composite lesions. A pragmatic criterion for division into lesion groups was used, and the lesions were, of course, not truly discrete. A statistical method was therefore used for partialing out the correlations between lesion sites within a defined range of lesions. The problem selected for this analysis was the question of the role of the insula and the basal ganglia in the context of varying lesions of the Broca and Wernicke areas. This problem was previously addressed by Brunner, Kornhuber, Seemuller, Suger, and Wallesch (1982), who found that a basal ganglia lesion, in combination with a cortical lesion, results in a more severe aphasia than a pure basal ganglia lesion. The studies by Mohr (1976), already cited, concluded that the addition of an insular lesion to a Broca area lesion is critical for producing a Broca aphasia. This conclusion was also confirmed in our own analysis (Reinvang, 1983).

7.3.1. Method

CT-scan was performed on clinical indications at various hospitals in Norway. For interpretation of the scans, a qualified neurologist or neuroradiologist (Drs. P. Borenstein and G. Dugstad) rated the presence of lesions on a checklist prepared by the author (Table 7.1). On the basis of the checklist, a more limited set of lesion categories or lesion groups was derived (Table 7.2). If the scans were suited for detailed interpretation, then diagrams of the lesions were made on standardized slice diagrams. The system was developed in the Department of Radiology, Neuroradiology, and Neurology at Aachen, West Germany (Schmachtemberg, Hundgen, & Zeumer, 1983; Blunk et al., 1981). Diagrams representing 16 brain slices with 5-mm separation were prepared, based on corresponding anatomical cuts shown in Matsui and Hirano (1978, pp. 143–157). A grid coordinate system was superimposed on each slice. The grid system had 58 × 42 points and corresponded, according to the authors, to the degree of resolution that can realistically be achieved in CT-scans.

In the fully implemented system, the lesions were processed automatically after transfer to the grid model. In the present study, the grid model was used only as a standardized mapping system and as a means for estimating total lesion volume. The estimate was done by counting the grid squares encompassed by a lesion, added over slices. The mappings were performed without knowledge of the aphasia test results by a neurologist experienced in clinical aphasia research (Dr. P. Borenstein).

Table 7.1. Lesion Checklist and Grouping

Area checklist	Lesion category (L category)
Broca area	Broca
Insula Capsula externa	Insula
Wernicke area	Wernicke
Supramarginal gyrus Angular gyrus	Posterior
Frontomedial Temporomedial	Medial
Cingulate gyrus Hippocampus	Limbic
Basal ganglia Pulvinar Other thalamic	Subcortical nuclei
Arcuate fasciculus	Arcuate

7.3.1.1. Subjects. The population of 249 patients contained 125 individuals with CT-scans. Some of these patients were excluded because the scans were not suitable for interpretation because of technical problems or because they had been performed too soon after the onset of the illness to permit evaluation of lesion localization. We found 89 patients with interpretable scans. Six different hospitals made CT-scans available for this study.

Table 7.2. Derived Lesion Groups

	Lesion in	
	Broca area	Wernicke or posterior
Group 1	+	0
Group 2	0	+
Group 3	+	+
Group 4	0	0

7.3.2. Results

7.3.2.1. Predictive Value of Classification Systems. The program for canonical discriminant analysis of the SPSS (Nie *et al.*, 1975) was used to predict a fourfold classification (severe nonfluent, mild nonfluent, severe fluent, and mild fluent) from the list of lesion variables (L1–L8), with total lesion volume added as an extra variable.

The program first factor-analyzed the correlation matrix and then used the derived factors in a linear equation for prediction. Two procedures may be used, one in which all of the terms in the variable list are used, and one in which the program determines empirically how many variables are to be used. Both procedures were used, with little variation in results. The results reported are those found with the direct procedure (all lesion variables used).

In all, 66% of the patients were correctly classified (see Table 7.3). The program identified three groups of predictive lesion variables: The first function refers to Broca area lesions; the second to a combination of arcuate fasciculus, medial, and insular lesions; and the third is represented by Wernicke lesions and lesion volume.

The discriminative predictive factors pointed out by the procedure make sense in that the classical language areas clearly came out as discriminative variables.

In predicting aphasia type, the same statistical procedure was used, but only classifiable cases were included. It was regarded as appropriate to reduce the number of categories somewhat by combining transcortical motor, transcortical sensory, and isolation syndrome into one group. The classical model predicts that they will share the property of having lesions outside the classical language

Table 7.3. Prediction of Aphasia Group

	Predicted group				
	1	2	3	4	
1. Severe nonfluent	15	1	3	0	*n* = 19
2. Severe fluent	0	6	0	2	*n* = 8
3. Mild nonfluent	2	2	7	1	*n* = 12
4. Mild fluent	1	5	4	12	*n* = 22

areas. Jargon aphasia and Wernicke aphasia were also combined in a common category, as the motivation for separating them was not the expectation of differential lesion locus.

In all, 36 classifiable cases were included in the analysis.

In all, 66% were correctly classified (see Table 7.4). The program identified five functions used in prediction. The first function is represented by arcuate fasciculus, the second by insula and Broca area, the third by Wernicke area and posterior areas, the fourth by lesion volume and subcortical nuclei, and the fifth by medial lesions.

Again, the lesion variables singled out as discriminatory make sense in terms of classical models of language pathology.

Although strict classification criteria were used in order to improve on predictive validity at the cost of reduction in numbers, the percentage of correct classifications predicted by the lesion data is not impressive.

The discriminant analysis technique was also used to predict deviations in reading and writing from auditory-vocal functions. The previous analysis of these conditions (Chapter 4) shows a variety of combinations, of which several were not represented in the present sample. Some combinations of categories, therefore, had to be constructed. It was decided to run an analysis for predicting a threefold classification: relative deficit of reading or writing (alexia or agraphia), no deviation in reading or writing, and relative preservation of reading or writing ("hyperlexia" or hypergraphia). We analyzed 15 cases with relative deficit or preservation.

Of the cases, 53% are correctly classified (see Table 7.5). Because

Table 7.4. Prediction of Aphasia Type

	Predicted type						
	1	2	3	4	5	6	
1. Global	6	1	0	0	0	0	$n = 7$
2. Broca	1	7	1	1	0	0	$n = 10$
3. Wernicke	0	0	3	0	1	2	$n = 6$
4. Transcortical	0	0	1	2	1	0	$n = 4$
5. Conduction	0	1	0	1	5	0	$n = 7$
6. Anomic	0	0	1	0	0	1	$n = 2$

Table 7.5. Deviation in Reading or Writing

	Predicted group			
	1	2	3	
1. Alexia or agraphia	3	1	0	$n = 4$
2. No deviation	6	23	16	$n = 45$
3. Hyperlexia or hypergraphia	2	3	6	$n = 11$

of the relatively poor discrimination between groups, the content of the predictive functions will not be further discussed.

7.3.2.2. Discussion. In the previous analysis by Reinvang and Dugstad (1981) and Reinvang (1983), the conditional probabilities showed a good correspondence between some aphasia types and lesion groups. Global and Wernicke aphasia predicted lesion group well, other types less well. The present attempt to improve correspondence by limiting the analysis to classifiable cases and using a more high-powered statistical procedure to optimize predictive relationships did not succeed very well. Beyond confirming that the variables singled out in the classical clinicopathological models are of discriminatory value, the results seem to confirm the essentially probabilistic nature of the relationship between pathology and function.

7.3.3. Analysis of Test Parameters

In examining the different parameters of the aphasia test in relation to the locus of injury, two types of analyses are performed. In the first (Table 7.6), a t test was performed on the difference between the groups having or not having a lesion of any designated area.

The result of this analysis is that, with respect to fluency, the Broca area, the insular region, and the subcortical nuclei are significantly involved. In other parameters, all classical language areas plus the insula are involved. There are two modifications to this statement. One is that the insula is not involved in auditory comprehension. The other is that, with regard to repetition, the arcuate fasciculus has a special function.

The second analysis is correlational, showing the extent of covariation between the degree of lesion in any given area and the aphasia

Table 7.6. Results of T Tests Comparing Patients Having or Not Having a Given Lesion[a]

	Broca	Insular	Wernicke	Posterior	Medial	Limbic	Subcortical	Arcuate
Words per minute	−						+	
Auditory comprehension	+	+	+	+				
Repetition	+	+	−	−				+
Naming	−	−	−	−				
Reading comprehension	−	−	−	+				
Reading aloud	+	+	+	+				
Writing	+	+	+	+				
Aphasia coefficient	+	+	+	+				

[a] + = Significant deficit in groups having the designated lesion ($p < .05$). Empty areas indicate no significant difference.

test parameters. The total lesion volume was added to the list of anatomical categories. The pattern of results indicated by the *t* tests are, on the whole, confirmed. It should be noted that, in addition to correlation of lesion in any given area with different parameters, total lesion volume has a significant correlation with all test parameters. The arcuate fasciculus stands out more clearly in this analysis as intimately connected to the classical language areas in functioning (Table 7.7).

7.3.4. Lesions and Their Context

The above is a descriptive account of relationships in the present pool of data. Our problem in interpreting these data is that lesion sites are not independent of each other. The distribution of the blood supply of the brain determines the likelihood of particular combinations of lesions in vascular cases. Ideally one would like to make a parametric study of the effect of, for example, a Broca area lesion in isolation and in all possible permutations with lesions in related areas. This ideal can be reached only in an experimentally controlled design, and in clinical research, we are forced to use statistical means of trying to correct for biases (in this case, correlations between lesion sites) in the independent variables.

The first approach to this problem is to try to disentangle the seemingly uniformly high contribution of all classical language areas by excluding from the analysis all patients with combined lesions of Broca and Wernicke areas and all patients with no lesions of the classical language areas.

The *t* tests and correlation matrix for this remaining group are shown in Tables 7.8 and 7.9.

It is not reasonable to analyze medial or limbic lesions because these occur primarily in a context without lesions of classical language areas, and this lesion group was excluded from the analyses.

It is worth noticing that only a few significant correlations remain. The influence of lesions in the Broca and posterior areas, as well as the arcuate fasciculus, is no longer significant. The general lesion volume affected only auditory comprehension and reading comprehension. Patients with lesions of the Wernicke area showed a higher rate of speech output and lower comprehension than patients without a Wernicke lesion. Reduced speech rate was associated with lesions

Table 7.7. Correlations of Lesion Size in Given Area and Performance (N = 61)ᵃ

	Broca	Insular	Wernicke	Posterior	Medial	Limbic	Sub-cortical	Arcuate	Volume
Words per minute	−.28	+.38					+.33	+.28	−.34
Auditory comprehension	−.31		−.37	+.36				+.30	+.49
Repetition	−.34	+.29	+.32	+.32				+.36	+.39
Naming	+.42	−.21	+.39	+.36				+.32	+.42
Reading comprehension	−.38		−.20	−.32				−.29	+.55
Reading aloud	+.42	+.34	−.20	−.32				+.31	+.40
Writing	+.47	+.20	−.18	−.25		+.20		+.32	−.45
Aphasia coefficient	+.42	+.23	−.35	+.38				+.37	+.52

ᵃEmpty areas have insignificant correlations.

Table 7.8. Results of T Tests Comparing Patients with Restricted Lesions of a Given Area[a]

	Broca	Insular	Wernicke	Posterior	Subcortical	Arcuate
Words per minute			[b]			[c]
Auditory comprehension						[c]
Repetition						
Naming						
Reading comprehension						
Reading aloud						
Writing						
Aphasia coefficient						

[a]Empty areas mean no significant difference between groups with and without lesion.
[b]Means that larger lesion is associated with higher score.
[c]Means that larger lesion is associated with lower score.

of the insular region and of the subcortical central nuclei (basal ganglia or thalamus).

The hypothesis immediately suggests itself that the richer set of correlations in the total group is a spurious effect of correlations between the independent variables. This is certainly true, but not sufficient to explain the findings.

Consider a task like naming. According to the analysis of restricted lesions, naming performance is unpredictable from either lesion locus or lesion volume. In the total group, both Broca, Wernicke, posterior, and arcuate lesions, as well as lesion volume, predicted naming quite well. If naming is related to a limited locus, then it should have shown up in the analysis of cases with restricted lesions. If naming is related only to lesion volume, then the volume effect should have shown up even in the restricted-lesion cases, among whom many had large lesions (mean lesion volume 1,469 units, range 376–5,573 against 1,474 units as the mean for the whole sample).

It may thus be that the relation between function and localization is different in restricted and composite lesions, but to test this possibility, a more rigorous analysis must be performed.

The problem of analyzing the contribution of insular or basal ganglia lesions in the context of lesions of the Broca area, the Wernicke area, or both was addressed. Three groups were formed. The first group (GA) had Broca area lesions or lesions outside the classical language areas. The second group (GB) had Wernicke area lesions or lesions outside the classical language areas. The third group (GC) had combined Broca and Wernicke area lesions or lesions outside the classical language areas.

The program "regression" from SPSS (Nie *et al.*, 1975) was performed in the 3 groups separately. The previous tables indicated that words per minute and auditory comprehension had a relationship to some of the areas considered in this analysis. The multiple regression took the correlation of a Broca area lesion with the variable in question. It added the partial correlation of a Wernicke area lesion to the same variable and proceeded with the insular region and the basal ganglia.

Figure 7.1 shows the cumulative curve of percentage variance explained for fluency by each lesion in this procedure. The significance of each partial correlation was tested with an F test.

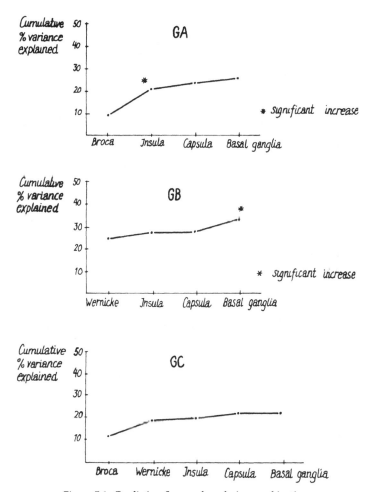

Figure 7.1. Predicting fluency from lesion combinations.

The equivalent function for comprehension is shown as Figure 7.2. The figure demonstrates that an insular lesion partially determines fluency when there is a restricted Broca area lesion. It does not have the same influence in the presence of a Wernicke area lesion or a combined Broca and Wernicke lesion.

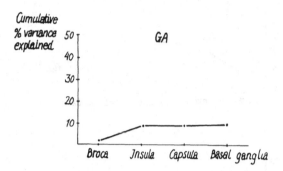

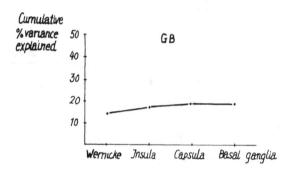

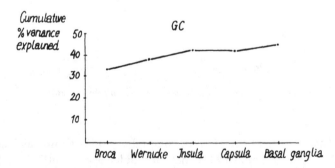

Figure 7.2. Predicting auditory comprehension from lesion combinations.

Table 7.9. Correlations of Lesion Size in Given Area and Performance in Patients with Restricted Lesions (N = 28)[a]

	Broca	Insular	Wernicke	Posterior	Subcortical	Arcuate	Volume
Words per minute		÷.37			÷.35		÷.36
Auditory comprehension			÷.31				
Repetition							
Naming							
Reading comprehension							÷.45
Reading aloud							
Writing							
Aphasia coefficient							

[a] Empty areas have nonsignificant correlations.

Auditory comprehension can be explained only in a combined Broca and Wernicke lesion. The apparent specific effect of a Wernicke area lesion in Table 7.9 seems to be a spurious effect, and the true effect is a volume effect in the context of classical language area lesions.

7.4. Conclusion

It is difficult to predict the type of aphasia from combinations of lesion variables. The parameters singled out by classical clinico-pathological models as distinctive also emerged in these statistical analyses, but not with predictive relationships of highly significant strength. A possible explanation of these probabilistic relationships can be derived from the analyses of restricted and composite lesions. At least in some cases, it can be shown that the effect of a given lesion is different according to the context of additional lesions. This inter-action will obscure any pattern of results pursued with methods look-ing only for independent effects.

The results are thus in favor of the third position outlined initially in this chapter. A concept of interaction- or context-dependent lesion effects is needed to supplement a classical localizationist form of anal-ysis. The results are in accord with reports by neurosurgeons that limited cortical removals give mild deficits (Penfield & Roberts, 1959) and that combined cortical and subcortical involvement predicts a more severe deficit (Hécaen & Consoli, 1973).

RECOVERY AND
PROGNOSIS

8.1. *The Recovery Process*

As argued by Sarno (1976), knowledge of the natural process of recovery is an important prerequisite for judging the efficacy of therapeutic efforts. In Chapter 1, it was also suggested that such knowledge has theoretical importance because, by comparing the immediate and long-term adjustment of the brain to a structural injury, the mechanism of localization of function in the brain (systemic or nonsystemic) may be assessed. The tendency to ignore change because it introduces noise into a system for predicting the locus of the lesion from aphasia type is unfortunate.

Just what the conditions are under which a natural process of evolution can be assumed to take place is unclear. It has been pointed out that any confounding influence of therapy should be eliminated if spontaneous recovery is to be assessed (Sarno, 1976). Reinvang and Engvik (1980a) and others have pointed out that there is an equally large risk of confounding influence by negative factors of deprivation and isolation. The more slowly developing recovery processes at work beyond the first few weeks after stroke or trauma must operate in a context of intrapsychic and social influences. We have suggested that "neutral treatment" is an appropriate control condition, and we define it as

and he is encouraged to make maximum use of intact verbal capabilities
for communication. (Reinvang & Engvik, 1980a, p. 79)

Previous studies describing general recovery trends have suggested that, in untreated patients, spontaneous recovery is seen only
for the first 2 to 3 months (Vignolo, 1964; Culton, 1969). This suggestion was contradicted by Kertesz and McCabe (1977), who found
continued improvement, but at a decelerating rate, for the first year
in unrehabilitated aphasics.

In patients having received some form of treatment, a longer
duration and a greater extent of recovery have been found, perhaps
dependent on the duration of treatment and the time of starting treatment (Vignolo, 1964; Basso et al., 1975, 1979).

The question of whether all aspects of aphasia improve to the
same degree is somewhat controversial. The most frequent opinion
is that auditory comprehension improves more than expressive performances (Vignolo, 1964; Basso et al., 1979; Lomas & Kertesz, 1978;
Prins, Snow, & Wagenaar, 1978). However, some authors have found
better recovery of repetition (Kenin & Swisher, 1972) or naming
(Kreindler & Fradis, 1968; Reinvang & Engvik, 1980a). The disagree-
ments may be due to differential recovery rates for different functions,
or different degree of recovery in subgroups. The latter explanation
was suggested by Lomas and Kertesz (1978).

The question of "Syndromenwandel" (Leischner, 1972)—that is,
how often the aphasia type may change—has rarely been discussed.
Kertesz and McCabe (1977) found relative stability of aphasia types
over time. Reinvang and Engvik (1980a) found that aphasia types
were, on the whole, rather stable because changes with time were of
a general nature and did not alter the shape of the test profiles.
Changes of test profiles leading to redefinition of the aphasia type
did occur, however. This occurrence has also been noted by others
and has led to questioning whether all of the traditional aphasia syndromes should be regarded as independent entities. Mohr, Pessin,
Finkelstein, Finkelstein, Duncan, Davis, and Grand (1978) regarded
Broca aphasia as being the result of improvement in global aphasia.
Several other authors, among them Liepmann (1915), have recognized
this development, but it is not commonly accepted as the only context
in which Broca aphasia can occur.

Doubt has also been raised about the independent status of

Doubt has also been raised about the independent status of conduction aphasia. Some authors think that it may be related to Wernicke aphasia, as a stage in the process of recovery (Kertesz & Benson, 1970). On the other hand, Benson et al. (1973) reported that conduction aphasia occurs frequently in the first weeks after brain injury, with subsequent recovery.

Anomic aphasia is regarded by several authors (Goodglass & Kaplan, 1972; Kertesz & McCabe, 1977) as an end point of development for improving aphasics with various aphasia types, but it may also appear as a primary syndrome.

The studies cited above lead one to expect that, by specifying the time interval during which improvement is studied, the type of task, and the type of aphasia, some of the ambiguities in the reported findings can be resolved.

8.2. Recovery of Nonverbal Functions

The recovery of nonverbal functions has been little studied, but some information was given by Kertesz (1979). In general, he found a greater recovery of nonverbal functions than of language functions in global aphasics, whereas in other types of aphasia, the nonverbal recovery was said to lag behind. The conclusions were based on visual inspection of test–retest results, with no control for level of initial performance.

Poeck (1983a) asserted that aphasia and ideational apraxia have different recovery rates and, on that basis, argued for separate underlying mechanisms. Kertesz (1979), on the other hand, reported a close parallel between recovery from apraxia and language recovery.

8.3. Prognosis

The prognosis for complete recovery from aphasia is poor in patients in whom the symptoms persist beyond the first few weeks (Culton, 1969; Brust et al., 1976). Prognosis in the present context is therefore a question of predicting the amount of improvement in individuals who will remain aphasic. Some background factors are believed to be of predictive significance. Several authors agree that etiology (traumatic vs. vascular) is of importance and that patients with traumatic etiology have the best prognosis (Butfield & Zangwill, 1946; Kertesz & McCabe, 1977). Age is believed to be important, with young patients making the best recovery. This trend, however, failed to reach statistical significance in several studies (Kertesz & McCabe, 1977; Sarno & Levita, 1971; Smith, Champoux, Levi, London, & Muraski, 1972).

There is no direct evidence that sex or education is prognostically significant, although McGlone (1980) speculated that the lower incidence of women reported in many samples of aphasics may be due to better prognosis as a result of the more diffuse cerebral representation of language in women.

Kertesz and Sheppard (1981) explained the reported differences in sex distribution as being due to the differential risk of stroke in the age groups from which candidates for study are recruited. They also failed to show differential recovery rates in the two sexes. Kimura (1983) found that differences between the sexes appeared with lesions of the anterior language area. These were also found by Fredriksen and Lernæs (1984) in our group.

Left-handedness has been stated to be associated with a better prognosis than right-handedness (Subirana, 1969), but the evidence is equivocal.

With respect to subgroups, Kertesz and McCabe (1977) found that Broca aphasics showed the best improvement and globals the poorest among all types of aphasics. These were untreated patients. Sarno, Silverman, and Sands (1970) found no improvement in globals even with treatment. However, Sarno and Levita (1979) reported some improvement in globals in the latter part of the first poststroke year. This finding was confirmed by Sarno and Levita (1981). They stressed the fact that these patients were "alert" globals. Mohr et al. (1973) found continued improvement long after injury in a case study of global aphasics. Wernicke (1874) regarded sensory aphasia (Wernicke aphasia) as having a good prognosis. Basso et al. (1975, 1979) found the improvement in fluent aphasics to be similar to that in nonfluent aphasics. Lomas and Kertesz (1978) regarded the initial level of auditory comprehension as a good predictor of improvement in other language performances. Brust et al. (1976) found that fluency was a good predictor of spontaneous recovery, with fluent patients improving most, whereas Vignolo (1964) reported a poor prognosis for patients with severe oral expressive (apraxic) problems.

8.4. Mechanisms of Recovery

Recovery has both a functional and a physiological aspect, and a satisfactory theory accounts for both. A predominantly neuropsychological theory need not specify the nature of the physiological mechanisms involved in recovery. A neuropsychological theory should specify the areas of functioning involved in recovery, the influence of premorbid factors, the environmental influences in the recovery period and the form of organized interplay of functional processes taking place. The findings of cellular mechanisms of recovery, both anatomical (Raisman & Field, 1973; Schneider, 1973) and biochemical (Glick & Zimmerberg, 1978), are valuable, but they do not constitute an alternative to neuropsychological models, nor do they exclude any specific form of neuropsychological model, with the exception of models that explicitly state that all recovery is an epiphenomenon based on compensatory mechanisms for covering up the deficit.

In the following, some theoretical alternatives for neuropsychological recovery models are outlined. They are ordered along a dimension introduced in Chapter 1, the dimension of degree of systemic relationship between brain regions. The simplest theory states that recovery is dependent on internal processes in preserved parts of the language areas. The most complex theory sees recovery as one aspect of readjustment of the whole brain to localized injury. Two in-between positions are outlined.

8.4.1. Relearning or Facilitation

Simple relearning by selective practice of language content material (words and sentences) was found by Wiegel-Crump and Koenigsknecht (1973) and by Wiegel-Crump (1976). Learning generalized to language material of similar content or structure. Schuell (1974) has been a strong advocate of using learning principles in therapy.

Facilitation is invoked by authors favoring a psychosocial stimulation treatment of aphasia (Wepman, 1953). The aphasic suffers
mental blocks and frustrations. If they are removed by appropriate
therapeutic attitudes, the way lies open for easier access to previously
learned language material.

Sometimes, aphasics experience a sudden breakthrough or rapid
improvement of speech as a result of strong emotional stimuli. Lifting
of inhibition has been invoked to explain such phenomena, which
mainly occur in the early weeks of the recovery process (Luria, 1970).

It is reasonable to view both simple relearning and facilitation
as a function of increased efficiency in the injured structures normally
responsible for the function. Authors favoring a learning approach
are typically not concerned with the neural basis of performance.

8.4.2. Reorganization of Function

Reorganization of function (Luria, 1966) is a complex notion
based on the idea that performances rest on functional systems. Injury
interferes with functional systems, but by reorganizing the remaining
components or adding new ones, a new foundation for adequate
performance is created. Dressed in terms of cognitive theory, one
may say that the patient learns or spontaneously invents new cognitive strategies. Some studies (Weinberg, Diller, Gordon, Gerstman,
Lieberman, Lakin, Hodges, & Ezrachi, 1977) with nonaphasic stroke
patients have shown that it is possible to teach patients strategies for
directing attention by making an intact resource (verbalization) the
key element in the strategy. Attempts to teach memory strategies to
aphasics by encouraging visualization have been only moderately
successful. These approaches differ from relearning approaches mostly
in terms of focus. Relearning focuses on concrete content material
(word and sentences), whereas reorganization focuses on general
strategies. It is also explicitly stated that the neurological basis for the
reorganized function is partly different from the normal basis.

8.4.3. Release of Vicarious Neural Structures and Functional
Relocalization

It seems obvious that the recovery process must be based on
activity in preserved brain structures. Either the right hemisphere
takes over, or preserved areas of the left hemisphere are responsible

for recovery of function, or both. Smith *et al.* (1972), finding significant correlations between improvement and a wide range of phenomena, suggested that wide areas of the brain, even those traditionally termed sensorimotor, take part in recovery. Milner, Branch, and Rasmussen (1964) concluded, on the basis of studies of cortical ablations in epileptics with injury dating back to childhood, that dominance for the language function does not shift unless there is a massive left-hemisphere injury. Rasmussen and Milner (1975), however, speculated that dominance for the anterior or posterior speech center may shift independently, indicating the possibility of a partial shift of dominance with less severe injuries. These hypotheses indicate that different recovery mechanisms are at work in severe and in less severe aphasics (right and left hemisphere, respectively). When relocalization takes place, a release from inhibition permits dormant functional competence to be utilized.

The hypothesis that right-hemisphere mechanisms are involved in recovery has been tested on aphasics with the dichotic listening technique (Johnson, Sommers, & Weidner, 1977; Pettit & Noll, 1979; Castro-Caldas & Botelho, 1980). All authors found some evidence for increased right-hemisphere participation in recovery, but Castro-Caldas and Botelho (1980) found the evidence only in the case of fluent aphasics. The differences on which these conclusions are based are small, and the results vary with the nature of the test.

8.4.4. Complementary Redifferentiation of Function

The hypothesis of complementary specialization has been applied to the cerebral hemispheres and implies that the specialization of one hemisphere is linked to the specialization of the other. This hypothesis was reviewed in Chapter 1. It may be applied to recovery by viewing recovery as partly a product of redifferentiation of function in a preserved neurological substrate.

A paradigm that may be particularly useful in shedding light on the sort of complex interactive process considered here is the paradigm of the single versus the serial lesion. A lesion of an anatomical structure generally has more serious consequences when it occurs in one stage than when it occurs in several stages with time intervals between them. Even when both groups of animals recover, lesions in additional structures have different effects in the two groups. The available evidence was reviewed by Finger (1978). Although several explanatory principles must still be considered, it seems that simpler mechanisms like neural shock effects cannot account for the data. A model of functional reorganization taking into account the systemic properties of interconnected neural structures is needed (Finger, 1978).

8.5. The Present Study

The present study asks three questions. They are answered and discussed in separate sections:

1. What is the recovery pattern for verbal and nonverbal per-formances? This question has both quantitative and qualita-tive aspects. The question of stability of clinical aphasia syndromes will be addressed.
2. Which are the important prognostic indicators? In addition to the language and neuropsychological variables, back-ground variables of age, sex, education, etiology, and lesion volume are considered.
3. What is the structure of relations between functions in recovered and unrecovered patients?

The answers to these three questions, taken together, are per-tinent to deciding between the neuropsychological models of recovery mechanisms. Although they are not precisely formulated, Table 8.1 is an attempt to summarize which findings to expect on the basis of different hypothetical recovery mechanisms.

A full discussion of the results with respect to this table will not be attempted here. How we view mechanisms of recovery depends on how we view cerebral localization of function, and this discussion, built on an integration of material from the different chapters, is reserved for the final chapter.

8.5.1. Recovery Pattern

In all, 134 patients were retested. The likelihood of being retested was compared in the different groups of the $2 \times 2 \times 2$ "cube" model (Chapter 3) and was not found to be significantly biased.

Table 8.1. Hypothetical Recovery Mechanisms and Findings Consistent with Them

Behavior mechanism	Neurological substrate	Recovery pattern	Functional structure	Prognostic signs
Effectivization Facilitation Simple relearning	Partially injured language areas	No nonverbal deficit. Global recovery of all impaired functions at decelerating rate.	Performance structure unaltered by time or recovery.	Initial state of language function. Time after injury. Lesions volume. Age.
Reorganization New strategies	Whole brain	Quantitative and qualitative jumps. No change in nonverbal abilities.	Change toward closer integration of verbal and nonverbal function in recovered patient.	Appropriate training. Preserved neuropsychological functions. Negative effect of lesion volume. Age.
Vicarious function	Right-hemisphere homologue	Nonverbal function may suffer if it competes with verbal function. Best recovery of comprehension and single-word utterances.	Relations of verbal and nonverbal performances unaffected for simpler tasks but may compete in more resource-demanding tasks.	Better recovery with large than with medium-sized lesions.
Complementary redifferentition	Whole brain	Impairment of nonverbal function. Verbal and nonverbal recovery are coordinated.	Verbal and nonverbal function become more clearly differentiated in recovery.	*Negative* influence of lesion volume. *Positive* effect of lesion momentum.

Table 8.2. Stability of Classification

Test 2	Nonfluent		Fluent	
	Severe	Mild	Severe	Mild
Test 1				
F-S	0	1	10	8
F-M	1	1	0	37
NF-S	36	12	2	3
NF-M	0	17	0	7
Total	37	30	12	55

The likelihood of retaining the same classification at retest is shown both for the 4-group classification and the 11-group classification (Tables 8.2 and 8.3). Stability is present in both systems in the sense that 64% to 75% of the patients retained the same classification at retest.

When one studies recovery as a function of time, the possible confounding influence of other variables must be considered. The

Table 8.3. Stability of Aphasia Types

Test 1/Test 2	1	2	3	4	5	6	7	8	9	10	11	
1. Global	13									3	1	
2. Isolation		0								1	1	
3. Transcortical motor			0							1		
4. Broca				10						1	2	
5. Jargon					3				2			
6. Wernicke						3			1	1	1	
7. Transcortical sensory							3				2	
8. Anomic								1		2		
9. Conduction				1		1			10		3	
10. Mixed nonfluent			1	11					2	28	4	
11. Others						1			2	4	17	
Total	13	0	1	22	3	5	3	1	17	41	31	137

hypothesis that recovery is a function of the initial state was discussed with respect to a subset of the present study group by Reinvang (1983).

There were high test–retest correlations for the subscores of the aphasia test, but if the difference score (Test 2–Test 1) is taken as a dependent measure, then the correlation with initial score (Test 1) was not significant.

Another surprising finding (Reinvang, 1983) is that the test–retest interval was not correlated with the recovery score. The computations were rechecked on the present sample, with the same results.

Based on these findings, it is justified to show recovery as a function of time between the onset of aphasia and the initial test. Time was divided into 1-month intervals. Thus, the numbers in some groups were low, and the appearance of the curve in some sectors may be an unreliable indication of trends. One-way analyses of variance were performed across time groups, and the presence of linear or nonlinear trends was tested for.

As can be seen (Table 8.4), the trends were similar for different parts of the aphasia test, and all tests showed a significant time-related trend. The recovery of the aphasia coefficient is shown (Figure 8.1) to illustrate the findings.

8.5.1.1. Recovery of Nonverbal Functions. Memory functions were discussed in Chapter 5, and the factorial structure of the tests was discussed. A multifactorial structure was found. Although several of these factors were related to type or severity of aphasia, they did not show the time-related trends shown by the aphasia test variables. A

Table 8.4. *Recovery Trend with Time for Subtest of the NGA*

	Between-group		Linear trend		Nonlinear trend	
	F	p	F	p	F	p
Words per minute	2.44	.05	1.25	n.s.	2.58	.05
Auditory comprehension	1.92	.05	4.99	.05	1.61	n.s.
Repetition	1.96	.05	6.23	.05	1.53	n.s.
Naming	3.09	.01	8.44	.01	2.55	0.1
Reading comprehension	2.28	.05	7.56	.01	1.75	n.s.
Reading aloud	2.79	.01	15.09	.01	1.56	n.s.
Writing	2.66	.01	9.45	.01	1.98	.04
Aphasia coefficient	4.19	.0001	15.09	.001	3.10	.01

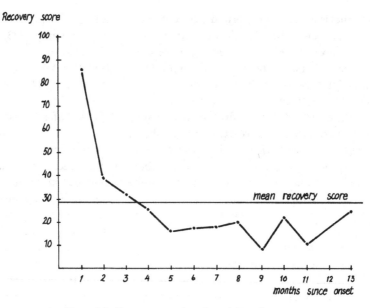

Figure 8.1. Recovery as a function of time since onset.

weak statistical trend was shown by two tests (DSE-I and BSE-I), but they gave too few clues to offer an interpretation.

The results support the position that the varieties of memory defects are not secondary consequences of the aphasic syndromes, but independent deficits.

Nonverbal abilities were discussed in Chapter 6, and the factor analysis showed two types of functions called *visual nonverbal abilities* and *apraxia*. The results on apraxia are clear-cut. The recovery did not parallel that of aphasia, and the results support the position of Poeck (1983a) rather than that of Kertesz (1979).

The factor analysis yielding a visual nonverbal factor could not distinguish functions measured by the Raven CPM from those measured by the Wechsler PIQ. The two tests behaved differently in recovery, and it was the Wechsler PIQ that seemed to show the same time-dependent phenomena as the aphasia coefficient. Other tests representing measures of apraxia, constructional deficit, or left-hand functioning showed no time-related recovery trend.

8.5.2. Prognosis

The aphasia coefficient difference score was used as measure of recovery. Prognosis was tested first with respect to the cube-classification system, with age added as a fourth dimension. A dichotomy was obtained by setting the cutoff point at the medium age of 53 years. The results (Table 8.5) show several independent effects.

Whereas higher order interactions could not be tested, there were four significant two-way interactions (Figure 8.2).

The interaction of fluency and chronicity was that the fluent and chronic group recovered less. The fluency and age interaction was that young nonfluents recovered less than young fluents, but old nonfluents recovered more than old fluents. This was a true crossover interaction. The severity and chronicity interaction was that severe patients recovered better than mild only in the acute phase. Recovery was less dependent on chronicity in mild patients. The severity and age interaction was that young and severe patients made a better recovery than any other group defined with these two variables. The finding of Sarno and Levita (1979) of better late recovery in nonfluent than in fluent patients was confirmed.

8.5.2.1. *Background Variables.* Age was analyzed in conjunction with aphasia variables because of the statistical association of age and severity, which can thus be corrected for. The results for other background variables (sex, education, and diagnosis) are shown (Table 8.6).

The significant effect of education was that the group classified as *students* improved more than other groups. This was obviously a young group, and it may be assumed that the age factor accounted for the apparent effect of education.

Table 8.5. *Recovery as Function of Subgroup*

	$\bar{x}1$	$\bar{x}2$	F	p
Type (nonfluent vs. fluent)	29.4	25.9	1.43	n.s.
Severity (low aphasia coefficient vs. high)	35.8	19.7	14.27	.001
Chronicity (acute vs. chronic)	30.7	19.1	5.71	.02
Age (young vs. old)	32.3	20.4	9.31	.01

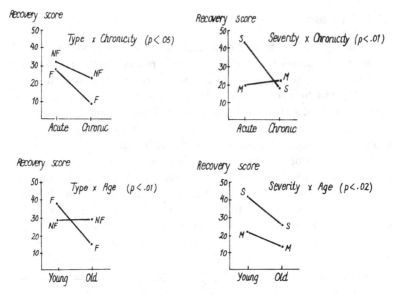

Figure 8.2. Two-way interaction effects.

Diagnosis was analyzed as thromboembolic, hemorrhagic (including subarachnoidal bleeds), head injuries, and others. Pairwise comparisons showed that the head-injured group improved more than the 2 cerebrovascular groups. To separate out the possible contribution of age to this finding was relevant but was not attempted. The problem was that there was little overlap in age in the cerebrovascular and the traumatic groups.

Table 8.6. Relationship of Aphasia
Coefficient-Recovery Score to Sex,
Education, and Diagnosis

Variable	df	F	p
Sex	1	.53	n.s.
Education	3	3.18	.05
Diagnosis	3	5.38	.01

8.5.2.2. Neuropsychological Variables. The set of neuropsychological variables analyzed in Chapters 5 and 6, measured at the first test, were correlated with the aphasia recovery score. No significant correlations were found in the group as a whole.

As recovery is sensitive to several dimensions of an aphasia classification, the analyses were repeated on subgroups suggested by variables showing significant main effects on recovery (Table 8.5). A summary of the results of these analyses is given (Table 8.7). The fluent versus nonfluent distinction is omitted because this distinction showed no main effect on recovery. The groups defined by the significant interaction terms were not analyzed because the number of subjects in each group was small.

Correlations derived by such an explorative procedure should be interpreted with reserve. If we bear in mind the factor analyses of neuropsychological tests, it is reasonable to require that several tests representative of a factor show significant correlations before this factor is taken to have a significant prognostic value.

Using the factor analyses as guidelines for interpretation, one

Table 8.7. Results of Correlating Neuropsychological Variables with Improvement in Aphasia Coefficient in Subgroups

Subgroup	Significant predictors of improvement
Severe	Pegboard, left hand ($-.26$)
	Digit span ($.31$); pointing span ($.20$)
	Apraxia tests FING ($.41$); MOV-I ($.40$); OBJ ($.29$)
	Block span ($.20$); block serial learning-sequence ($-.20$)
Mild	Raven CPM ($.24$)
	Frostig ($.50$)
	Apraxia tests MOV-I ($-.44$) and OBJ ($-.28$)
	Block serial learning-trials ($-.32$)
	Block serial learning-intrusion ($-.20$)
Acute	Digit span ($-.27$)
	Digit serial learning-perseveration ($-.22$)
	Apraxia test MOV-I ($.23$)
	Block serial learning-sequence ($-.19$)
Chronic	Finger tapping, left hand ($-.32$)
	Apraxia tests COPY ($-.39$)

Using the factor analyses as guidelines for interpretation, one may conclude that apraxia was a prognostic sign in both severe and mild aphasics. The prognostic relation had the opposite sign in the 2 groups, and therefore the effects canceled each other out in an overall analysis.

Verbal immediate memory (represented by two tests) was also a prognostic sign in severe aphasia. Combinations of nonverbal learning factors that could be recognized from the factor analyses showed up as significant in both severe and mild aphasics.

The clearest prognostic implications of neuropsychological functions appeared in relation to the severe-mild distinction. The findings are not so readily interpretable with respect to the acute versus the chronic distinction. Low but significant correlations on a scattered selection of tests are difficult to assess. In the following section, the underlying functions in acute and chronic patients are analyzed with another statistical approach.

8.5.3. Relations between Functions in Acute and Chronic Patients

The theoretical models outlined (Table 8.1) can be viewed as making different predictions about the structure of recovered language functions. The reorganization theory says that language becomes more closely interwoven with nonlanguage abilities because of compensatory strategies relying on intact abilities. The differentiation theory says that language and nonverbal abilities become less interdependent in recovery because of an underlying process of ongoing differentiation.

I have used the method of factor analysis (principal factor with varimax rotation) to assess the number of significant factors present in acute and chronic groups and the specificity of these factors. These notions are hard to give a satisfactory operationalization. The SPSS program for principal component analysis limits the number of factors analyzed by requiring eigenvalues at or above 1.0. The percentage of variance explained by each factor is also calculated. As a rule of decision, I will say that Functional Domain 1 was more highly differentiated than Functional Domain 2 if the factor analysis isolated a higher number of factors and if the proportion of variance explained by the first factor was lower for Domain 1 than for Domain 2.

Bear in mind that the recovery patterns may have been specific to subgroups. The factor analyses were performed on the subgroups suggested by the main effects in Table 8.5. According to this analysis, severity and chronicity were significant variables in addition to age.

The variables entered into the analyses were the neuropsychological tests described in Chapters 5 and 6 and the subtests of the aphasia test (words per minute was taken as measure of fluency).

According to the criterion suggested, it was the severe aphasic group that showed the clearest trend in the direction of greater differentiation of function with time. Six factors are pointed out in the acute group and seven in the chronic. The percentage of variance explained by the first factor shows a 5% decrease from acute to chronic patients (see Table 8.8).

8.5.4. Conclusions

The findings of the last two sections suggest strongly that the general recovery pattern described in the first section was not representative of every subgroup. Likewise, the general finding that neuropsychological variables show no close relationship to aphasia must

Table 8.8. Factors Revealed in Principal Component Analysis of Aphasia Test and Neuropsychological Variables

Group	Number of factors	Percentage of variance explained by first factor
All acute	4	45.5
All chronic	5	45.1
Acute: Severe	6	36.5
Mild	6	30.5
Young	5	45.8
Old	6	45.2
Chronic: Severe	7	31.4
Mild	7	29.8
Young	5	41.2
Old	5	44.0

be qualified.

The finding of distinctive patterns over time and across functions may indicate different recovery mechanisms in different subgroups. Because the differences between groups were not dramatic, it is more reasonable to conclude that several recovery mechanisms may have operated simultaneously, but with different weights in different groups.

The group yielding the clearest signs of a comprehensive adjustment of the whole brain to the injury was the severe aphasics. This was a group that made a greater-than-average recovery, especially if, in addition to being severe, they were in the acute stage and young. Several neuropsychological variables representing factors of apraxia, verbal memory, and nonverbal memory were predictive of recovery. Finally, the severe aphasics showed signs that a process of functional differentiation was taking place between the acute and the chronic stages of the illness.

9

THE ORGANIZED
RESPONSE OF THE
BRAIN TO INJURY

The assertion that the organization of the brain is complex is trivial, as is the statement that models reflecting this complexity must have more sophisticated options for explaining integrative and selective action than those of postulating the addition or the subtraction of independently localized functional components.

The clinical models of localizationist thinking seem to attribute to the brain less complexity of organization than general physiology would attribute to the normal human and animal brain. Although we suspect that the clinical models are too simple, we must know exactly in what respect they fail in order to introduce more complex models constrained by facts. This does not mean that all simple explanatory models must be abandoned. It may be that a model is applicable in a narrow domain although it is not applicable in a wider domain. It would constitute a scientific advance if a more generally applicable model could be proposed, in which the predictions of the first model are retained as special cases rather than general explanatory principles. The work of Wood (1978) is a good example. In his theory, the results of a localizationist model are incorporated as special cases in a more general associate-network model.

9.1. Evidence for Organized Complexity

A simple prerequisite for finding the evidence of complexity that motivates revisions of models is to look for it. The present study has looked for evidence of complexity by quantifying variables and using appropriate methods of experimental control and statistical analysis. Several examples of theoretically significant results can be given, of which three are the following:

Example 1. The procedure of using analogous tasks with verbal and nonverbal content was used in the analysis of memory. This experimental control, together with multivariate statistical methods, showed a complex pattern of organization that did not easily fit the verbal-nonverbal dichotomy. This dichotomy was not rejected. It was found to be a dominant but not exclusive mode of organizing the underlying functions.

Example 2. The degree of quantification of both lesion and aphasia variables permitted the search for an empirically derived regression function for predicting function from combinations of lesion variables. The procedure confirmed the discriminative value of lesions in the classical language areas, but not with impressive predictive success. Rather than stopping at this stage and reporting the results as a partial confirmation of the localizationist model, the further analysis of subgroups revealed the principle that the effects of lesions are context-sensitive (i.e., variable as a function of the presence of other lesions). There may even be areas (e.g., the basal ganglia) that are not language areas in the classical sense that an isolated lesion gives rise to aphasia. In the context of a lesion in classical language areas, these other areas may still contribute to symptom formation.

Example 3. The recovery curve shows a time-dependent decelerating shape and is relatively independent of associated neuropsychological findings. When multivariate methods are used to isolate subgroups with possible distinctive recovery patterns, the general view gives way to several contrasting patterns. The distinction between severe and mild aphasia seems essential to the study of the interdependence between aphasic recovery and associated neuropsychological function. In severe aphasia, this coupling seems most clear.

I hold that these three examples are sufficient to show organized complexity in clinical phenomena and their relation to injury with evidence of multiple interaction between factors. Thus, sufficient motivation, aside from plausibility outside the clinical context, is present for considering a systems theory type of explanation of the findings. In order to do so, I will offer a more general sketch of a model before applying it to explanations of clinical phenomena.

9.2. A Proposed Systemic Model

9.2.1. Abstract Model

Consider a set of tasks (1 . . . 6). Hypothetical Devices A and B are necessary and sufficient to perform the tasks, and when working conjointly as the system AB, they account for all the variance observed in the tasks.

Assume that we are able to test the functioning of A independently of B and vice versa. We find that A performs certain tasks better than B and performs some worse. Express this finding in terms of the percentage of total variance in a task accounted for by A and by B. The resulting curve shows the response characteristics of A and B and might look as in Figure 9.1.

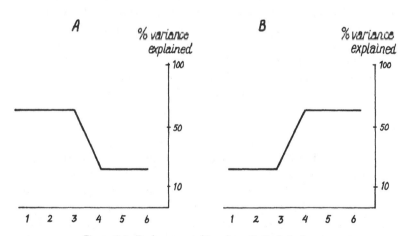

Figure 9.1. Performance of two hypothetical devices.

The devices have some degree of complementary specialization, in that A is relatively more proficient at solving 1, 2, and 3, and B is more proficient at solving 4, 5, and 6.

The variances of the two components R^2 (A) and R^2 (B) are related to the variance of the system R^2 (AB) by the formula for multiple correlation (Snedecor & Cochran, 1967, p. 402):

$$R^2 \text{ (AB)} = \frac{R_A^2 + R_B^2 - 2R_A \times R_B \times r_{AB}}{1 - r_{AB}^2}$$

In this equation, the term r_{AB} stands for the correlation of A with B. If the two devices are working independently, the correlation is zero and the variance of the output is the sum of individual variances, in this case about 85%.

If we want to improve on this performance, we might be able to improve on A by redesigning the components involved in solving Tasks 4, 5, and 6 so that they are specially designed to solve only Tasks 1, 2, and 3. That would be a good solution if Device A or its designer "knew" that Device B would always be at hand to take care of tasks 4, 5, and 6.

The above equation indicates another possibility. The correlation r_{AB}, if different from zero, will change the variance of the output without requiring a change in the individual components of A and B. It was postulated that the system (AB) could account for 100% of the variance in the tasks. It can easily be seen that a negative correlation would increase the total variance explained because it increases the numerator of the ratio. This means that activity in A reduces activity in B and vice versa. This relationship makes some sense if Devices A and B are coupled in such a way that they inform each other of what tasks they are best able to solve. As long as Device A is kept informed that another device with superior efficiency is operative, it does not do anything when Tasks 4, 5, and 6 are presented. By being freed of this responsibility, it is able to improve on its performance of Tasks 1, 2, and 3. If information comes that Device B is not working, then Device A reverts to the mode of operation shown in Figure 9.1. This mode of cooperation would be a truly systemic interaction, in which specialization is partly determined by the built-in structure of the device, but in addition, a dynamic component is added by way of a negative feedback coupling. One might say that the dynamic component serves to sharpen and enhance the performance profile of the device. Note that a positive correlation, in which activity in A stimulates activity in B, has the effect of making the output of (AB) less than the sum of the individual components.

9.2.2. Neural Model

It is suggested that, in the organization of higher nervous activity, principles of both structural specialization and dynamic sharpening of specialization are employed. Anatomically determined differences in response characteristics are referred to as *specialization*, and differences determined by dynamic interplay as *differentiation*.

Anatomical *specialization* means that cells with a special structure perform special functions. This mode of organization is present in the sensory cortex; the sensorimotor cortex, in which cells with closely similar response characteristics are localized in columns (Mountcastle, 1957) in the cortico-cortical connectivity, consists typically of short connections to neighboring cells (Jones & Powell, 1970).

Dynamically based *differentiation* of the response characteristic of a cell or a cell assembly is a common phenomenon at lower levels of the nervous system. In perceptual systems, lateral inhibition is a well-documented mechanism accounting for contrast enhancement and perceptual illusions like the Mach-band effect (Ratliffe, 1961).

Note that, at higher levels of the nervous system, differential functions may evolve in areas with similar cell structure. The most striking case is that of hemispheric specialization, in which homologous areas (i.e., areas with essentially the same structural characteristics) have different functions. In general, the so-called "association cortex" is present in several cerebral lobes, the cell groups are anatomically similar within this cortex, and differently located association cortices are connected by long fiber tracts. The functioning of these long fiber connections is crucial to the hypothesis that the presence of long connecting fibers allows distant cortical areas to develop *complementary differentiation*. Presumably, some structural asymmetry, in the form of either mild difference of cellular specialization or simple size difference between the connected areas, is necessary to set off a process of further differentiation.

The classical interpretation is that the long fiber tracts allow the formation of complex concepts based on association. In commenting on the functioning of callosal fibers, Pribram (1971) proposed a different interpretation:

> What kind of connectivity is it that rends asunder functions it supposedly associates? This question has not been asked until now. My own answer is that perhaps the connections, rather than functioning to associate, tend to separate through suppression the various parts of cerebral mantle. (p. 362)

The neural model is a further extension of this proposal. First, the evidence for the neural model will be examined, in relation to hemispheric specialization and differentiation, and then an application of the same model to anterior-posterior specialization and differentiation will be suggested. This is, to my knowledge, a novel application.

9.2.3. Clinical Evidence on Hemispheric Relationships

Evidence relating the functioning of the corpus callosum to the development of hemispheric lateralization was cited in Chapter 1 (Selnes, 1974; Denenberg, 1981). Some clinical evidence is relevant to further evaluation of the model.

A large brain injury changes the structure of the remaining tissue, as well as damaging fiber connections. Thus, the basis for both structural specialization and dynamic differentiation is altered. With large hemispheric lesions at an early age, the evidence suggests that lesions of the left hemisphere are followed by some deficits in both language and nonlanguage performances, but by no means by severe aphasias. This finding indicates that the right hemisphere is involved in restitution of the language functions in these cases, but that the loss of structurally specialized tissue cannot be fully compensated for.

Injuries to the right hemisphere are followed by normal language development and severe visuospatial deficits (Kohn & Dennis, 1974; Woods & Teuber, 1973). The higher recovery of hemisphere-specific functions in left- than in right-hemisphere lesions has led to notions that language development takes priority in development (Woods & Teuber, 1973). This notion is difficult to relate to neural mechanisms. How does a neurological structure "decide" to give up performing the functions that it does best? The present neural model offers an explanation.

Rather than saying that the preserved areas take over new functions, I hypothesize that neural tissue becomes functionally dedifferentiated when information about specialized activity in connected tissue is missing. It is interesting in this connection to note that the hypothesis can explain the asymmetry between hemispheres in the ability to compensate for early injury. Assume that the hypothetical Devices A and B correspond to the left and right cerebral hemispheres, and that Tasks 1 to 3 are verbal and 4 to 6 are nonverbal. As it stands in Figure 9.1, the model fails to predict the results because the hemispheres are depicted as having the same degree of structurally based specialization. The classical students of higher cortical function regarded the left hemisphere as "leading," or as specialized in relation to the right (Head, 1915), whereas modern students have tended to stress complementarity in the degree of specialization between hemispheres (Milner, 1974). Gazzaniga and Ledoux (1978) took up the thought that the only basic (structural?) specialization in higher cortical function is that of the left hemisphere for language, and that the apparent specialization of the right is a secondary effect of this (dynamically based differentiation, according to our terminology).

Assume that the relation between A and B in terms of structural specialization is as in Figure 9.2. The fact that A performs Tasks 1, 2, and 3 better and 4, 5, and 6 worse than B should lead to a complementary, but dynamically based, differentiation of response in B, so

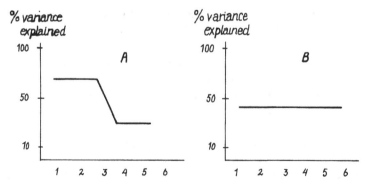

Figure 9.2. Performance of two devices, one of which is specialized.

that in a nervous system with matured connections between A and B, they would, in conjunction, operate as in Figure 9.1. If A received a massive injury, however, the basis for upholding the differentiation of B would be removed. B would revert to the mode of response determined in its inherent structure, that is, no advantage in responding to any specific type of stimuli.

A massive injury of A would give an even performance profile with somewhat subnormal performance on all tasks. A massive lesion of B would give an uneven profile with relative preservation of language performances. If A is taken as the left hemisphere and B as the right hemisphere, then this is exactly what is found in studies of early lesions to the hemispheres.

The adult lesion data concern mainly evidence for right-hemisphere function in recovery of language function. In some recovered aphasics, language function is interfered with when the right hemisphere is temporarily inactivated (Kinsbourne, 1971). The lack of knowledge of the preinjury degree of hemispheric specialization makes such studies somewhat difficult to interpret.

9.2.4. Within-Hemispheric Specialization and Differentiation in Humans

Consider the hypothesis that the anterior and posterior language areas function in analogy with Devices A and B in the conceptual model, and that the dense cortico-cortical connections between them (the arcuate fasciculus and possibly the uncinate fasciculus) are important in developing and to some degree maintaining differentiation of function. This hypothesis may be an aswer to the puzzlement expressed by Galaburda (1982) and cited in Chapter 2: How can richly interconnected and architectonically similar areas have different functions?

There is indication of incomplete anteroposterior differentiation of language areas in normal development, as reduction in fluency of speech follows injury in any part of the language areas in children, whereas it is related to anterior injuries in adults (Alajouanine & Lhermitte, 1965).

It may be discussed whether the model in Figure 9.1 or Figure 9.2 gives the best explanation of the data. The original Wernicke–Lichtheim hypothesis assumes that Figure 9.1 depicts the functional specialization of the language areas, with Device A specialized for articulatory programming and Device B for auditory language perception.

In the present study, the findings led to the acceptance of a less localized representation for most language performances, including auditory comprehension. The evidence for localization (specialization) is strongest with respect to fluency. Maintaining fluency depends not only on the Broca area but on the insula and possibly on the basal ganglia. If the term *anterior language area* is taken to refer to this anatomical subsystem, then I suggest that the relation between language areas is as in Figure 9.2, where A now is the anterior and B the posterior language area. As a result of further dynamic sharpening of specialization, the posterior language area will also perform in a differential mode, but depending to a greater extent on input from the anterior language area to uphold this differentiation than vice versa.

9.2.5. The Effect of Lesions and the Systemic Basis of Recovery

We have so far considered the situation where two components or areas of the brain are systemically related and serve to sharpen each other's response profiles by reciprocal inhibition. If one of the components is injured, then the other suffers some loss of ability to respond differentially. The loss may be moderate, as in Figure 9.1, or severe, as in Figure 9.2, depending on the design features of the preserved component. Loss of differentiation will be termed *Stage 1* in the response to injury.

Although one is apt to think in dichotomies of brain structures, as left-right or anterior-posterior, it is not likely the case that systemic organization is built only on structures connected in pairwise fashion. Figure 9.3 shows a more complex situation with three systemically related components.

Component A does best at Task 1, Component B at Task 2, and Component C at Task 3. The performance of these tasks is therefore "boosted" in the respective components, whereas the performance of other tasks is inhibited. The arrows indicate the direction of the systemic influence on the ability of a given component to perform a given task.

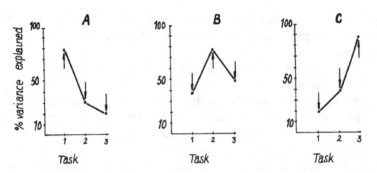

Figure 9.3. System with three components.

Injury to Component A causes some loss of differentiation in both Components B and C (Stage 1; see Figure 9.4). It now appears, however, that Component B is relatively better at Task 1 than Component C. This potential capability of Component B to solve Task 1 is "unmasked" by the lesion of A. The term *unmasking* has been used to describe the situation where neural pathways that are present but inefficient may assume functional importance when more efficient pathways are disrupted (for review, see Bach-y-Rita, 1981). The case

Stage 1

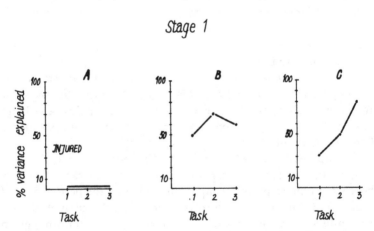

Figure 9.4. Loss of differentiation after injury (Stage 1).

described above is analogous to unmasking, but at a systems level. The same term has therefore been used, but in quotation marks.

With time, the reciprocal influence of B and C will lead to some redifferentiation of function in both components. This redifferentiation is termed *Stage 2* in response to injury. The most important consequence of redifferentiation in the hypothetical example is that, in Component B, its natural capacity for performing Task 1 is boosted rather than inhibited, as was the case before injury to A (see Figure 9.5).

An interesting question is whether Tasks 2 and 3 will return to a normal functional level. My guess is that the loss of one major systemic component cannot be fully compensated for. In any case, it is important to note that Tasks 1, 2, and 3 all improve in parallel and as a result of the same underlying process of differentiation. Although recovery takes place in parallel, there will, of course, be differences in the level from which recovery starts and in the level ultimately attained. The idea that some functions have to deteriorate or be "sacrificed" so that others may improve is foreign to this model.

The immediate effects of a lesion followed by dedifferentiation are encompassing deficits and high correlations between a wide range of performances. The physiological responses of the brain in the acute

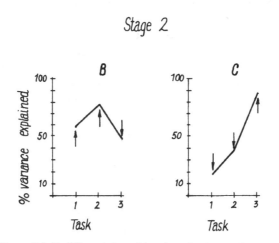

Figure 9.5. Redifferentiation of function after injury (Stage 2).

stage of illness tend to add to this picture, but the effects of dedifferentiation can be distinguished from those of acute physiological responses like edema by being in effect for a much longer time period, usually months after the insult.

With limited lesions, some specialized neurological structures are usually preserved, together with cortico-cortical pathways. This means that the basis of recovery may be a system in which little relocalization has taken place, if the preserved areas are able to uphold the previous pattern of localization.

9.3. Testing the Model

As stated in Chapter 1, general systems theory is not a testable theory but a framework for formulating theories to account for phenomena showing evidence of organized complexity. Adding specific assumptions makes a theory testable. Within a systems theory framework, the present theory has made several assertions that are in principle testable. These assertions may be summarized as follows:

Fully developed hemispheric specialization depends on a differentiation process. A necessary condition for such a process is an anatomical bias or asymmetry, which in the case of the hemispheres is the advantage in size of the language areas of the left side over the homologous areas of the right (Geschwind & Levitsky, 1968; Wada, Clark, & Hamm, 1975). Within the left hemisphere, anterior-posterior differentiation depends on a more rigid specification of functional characteristics of the anterior language area than of the posterior. The evidence for this conclusion comes from clinical data; however, there is as yet no anatomical evidence for asserting a higher degree of specialization of the anterior than of the posterior language area.

Areas that are connected via bidirectional fiber tracts, that show homological structure, and that show some degree of anatomical asymmetry are candidates for developing functional differentiation.

The nature of the input to such a system may serve to speed or retard the differentiation process. High demands for functional competence speed the differentiation process, and in general a more differentiated pattern of functional localization leads to higher functional efficiency.

The result of the differentiation process is not a conglomerate of single-channel processing devices, but a set of multichannel processing units differing from each other in functional profile rather than in the nature of the functions represented. Functional localization, even when resulting from dynamic differentiation, becomes more rigid and entrenched ("mechanized" is the term used in general systems theory) with advancing age. Structural damage to the system leads to some loss of differentiation but with some possibilities for new differentiation to develop.

9.3.1. Application of the Model to the Present Findings

In the introduction to this chapter, three examples are given of findings from the present study requiring a complex explanation.

In Example 1, the multiple overlapping organization of the functions underlying verbal and nonverbal memory is cited. The model described above is suited to explain this type of organization but does not contain assumptions that allow a further discussion of this example.

In Example 2, it is pointed out that the general features of the classical clinico-pathological model were confirmed by the results. These features are accounted for in the model by making the anterior-posterior difference in function a consequence of hard-wired specialization of the anterior language area. Strong indications were found of context-sensitive lesion effects, as would be predicted from the sort of interplay of dynamic forces postulated by the model. Because auditory comprehension is highly sensitive to an interplay between lesion loci, no anatomically specialized substrate for that function can be asserted.

If the present model is essentially correct, then some serious methodological consequences follow for clinico-pathological research. To determine the specific functional role of any anatomical structure is a monumental task requiring extensive parametric studies. Patients with "pure" lesions are not a sufficient basis for drawing any inferences.

In Example 3, several contrasting patterns of recovery are pointed out. This example is well suited to bring together several features of the model. *Severe aphasia* (large lesions) leads to more extensive involvement of nonverbal functions than *mild aphasia* (small lesions), because the cornerstones of functional differentiation are to a greater extent destroyed. In severe aphasics, the conditions favoring better recovery are those favoring a more radical loss of differentiation with ensuing redifferentiation and relocalization of function. If the lesion is anterior, then the posterior language area has a greater degree of freedom to enter into new relationships. If the lesion is posterior, then the anterior language area has less freedom to adjust because its mode of functioning is to a greater extent specialized. Add the higher degree of mechanization of function in advanced age, and the prediction follows that patients who are severe, fluent, and old have the worst prognosis. This three-way interaction could not be tested directly, but, of the possible two-way interactions (Severity × Type, Severity × Age, and Type × Age), the latter two were found to be significant. The Severity × Chronicity interaction showed that it was only the severe (not the mild) aphasics who made a better recovery in the acute than in the chronic phase. This interaction supports the hypothesis that there are different recovery mechanisms at work in severe and mild aphasics. In mild aphasics, the process underlying recovery is probably to a greater extent dependent on effectivization of processing in preserved parts of the language areas and development of effective cueing strategies. This process is not strongly dependent on time after injury.

9.3.2. General Applications of the Model

The model has the virtue of not postulating special mechanisms that only have application in a limited context. In addition to the model's applicability to research on hemispheric specialization, two examples of related models used to explain normal phenomena are mentioned here.

The phenomena of selective attention have been addressed by Kinsbourne & Hicks (1978), whose views were cited in Chapter 1. A paradigm for studying selective attention is the dual-task interaction paradigm, in which the person has to perform two tasks or processes at the same time. The results of such experiments bear on the nature of the cognitive processing system as a serial or parallel processing device, an issue of central concern in cognitive theory. There have been recent attempts to specify differentiated resource pools underlying cognitive performance and to tie them to a neurological substrate, that is, the cerebral hemispheres (Friedman & Polson, 1981;

Navon & Gopher, 1979). If it is hypothesized that cerebrally localized functions are resource pools, then the present model can explain both relatively constant, independent human resource pools and the ability of subjects to learn with extensive practice to perform some types of dual tasks by invoking the mechanisms of specialization and differentiation. By specifying necessary conditions for differentiation, the model avoids the emptiness of allowing unlimited and arbitrary postulation of resource pools. Walley and Weiden (1973) addressed the same problem area, taking the analogue of lateral inhibition in sensory systems (Ratliffe, 1961) as an explicit model for interaction and differentiation between cognitive functions.

The second example of a related model is found in the theory of Witkin, Goodenough, and Oltmann (1977), in which cognitive and emotional development is characterized as starting from a relatively global, diffuse, undifferentiated state and developing toward a more differentiated, context-independent mode of cognition. Witkin *et al.* (1977) cited cerebral asymmetry as a neural correlate of differentiation. The concept of *cognitive style* implies the ability of individuals to vary along a continuum of degrees of differentiation according to experiential background, sex, and situational demands. The concept of differentiation is also central in the theory of personality presented by Royce and Powell (1983). They presented evidence derived from factor analytic studies of a continuing developmental process of differentiation and hierarchization of functions.

The present model thus seems to incorporate concepts and processes central to the psychological theory of cognitive processes. Its novelty is in applying these concepts to the analysis of the consequences of cerebral injury. In doing so, it has prompted some viewpoints on the neurological substrate to emerge, which are of interest to cognitive theory.

9.4. Concluding Remarks

One of the starting points of the present essay was a plea for a research strategy that might yield results applicable to a wide range of aphasics, not only to selected and rare cases. It is appropriate, therefore, to devote the final discussion to this point.

The study has presented a lot of descriptive information that is generally of interest to clinicians and users of similar methods. More important, however, is the general model derived from these results.

The goal of cognitive and neurolinguistic analysis is to bring to light processing units that fail to function. The basis for inferring such units is a model based on the study of normal performance (Marshall 1982; see Chapter 1, p. 3). The underlying assumption is that the normal system can be used as a model for the impaired system. This assumption is of course essentially correct, but in the present view it is only strictly valid in the case of mildly aphasic individuals with small lesions. Only in these cases can the whole retraining program be based on the strategy of identifying, and then strengthening or bypassing, weak links in the chain of processing.

Our patients give us many clues that there is something more amiss with them than the failure to execute certain subroutines of information processing. They show various degrees of disability in conforming to task requirements; they are distractible, perseverating, or unable to refrain from unrelated activity (e.g., talking when a non-verbal response is called for). Some neuropsychological theories have tried to come to grips with such problems by stressing the preprocessing or "micro-genetic" aspect of any cognitive task. Before the highly differentiated information processing routine can be executed, the processing system has to go through stages of preparation to meet the task requirements.

The neuropsychological theory of Brown (1979) was cited earlier (Chapter 1, p. 15). He used the term "micro-genesis" and stressed the analogy between phylogenetic and ontogenetic development of a process and its unfolding in the present.

The neuropsychological theory advanced here differs from that of Brown in emphasis. The present theory does not stress hierarchical aspects of processing but emphasizes differentiation between types of processes dependent on cortical areas that are presumably closely related in both phylogenesis and ontogenesis.

An undifferentiated brain can only process integrated or "fused" information (e.g., by verbal-contextual-emotional integration). If the presented information is not fused but demands parallel or selective processing at an early (preprocessing) stage, then interference may result. Likewise, the emission of fused responses (e.g., by verbal-gestural-autonomic integration) may be feasible, but selective or sequentially organized activation is hampered by interference. These phenomena are well known to experienced clinicians, who can sometimes exploit them to make patients perform surprising feats of comprehension or expression. These severely aphasic patients, however, do poorly in formal training programs and on formal testing.

The fault with our approach to severe aphasics, then, may be that we think of them as only quantitatively different from the mild aphasics. The conclusion of this essay is that impaired brains differ in the degree of functional differentiation present after injury. If this conclusion is correct, both the content of therapy and the mode of task presentation in therapy must depend on assumptions about the capacity for differentiated processing present in the preserved system. A goal of therapy is not only to exploit the abnormal interactions of an undifferentiated system but to further the process of differentiation. A first step toward this goal is to understand the more general features of the differentiation process. Cognizant of the difficulties and of the many steps that will have to follow if practical benefits are to be attained, I offer the model presented here as a first step.

REFERENCES

Alajouanine, T. H., & Lhermitte, F. (1965). Acquired aphasia in children. *Brain, 88,* 653–662.

Alexander, M. P., & LoVerme, S. R. (1980). Aphasia after left hemispheric intracerebral hemorrhage. *Neurology, 30,* 1193–1202.

Alexander, M. P., & Schmitt, M. A. (1980). The aphasia syndrome of stroke in the left anterior cerebral artery territory. *Archives of Neurology, 37,* 97–100.

Arbib, M., & Caplan, D. (1979). Neurolinguistics must be computational. *The Behavioral and Brain Sciences, 2,* 449–483.

Arena, R., & Gainotti, G. (1978). Constructional apraxia and visuoperceptive disabilities in relation to laterality of cerebral lesions. *Cortex, 14,* 463–473.

Arrigoni, G., & de Renzi, E. (1964). Constructional apraxia and hemispheric locus of lesion. *Cortex, 1,* 170–197.

Atkinson, R. C., & Shiffrin, R. M. (1968). Human memory: A proposed system and its control processes. In K. W. Spence & J. T. Spence (Eds.), *The psychology of learning and motivation* (Vol. 2). New York: Academic Press.

Bach-y-Rita, P. (1981). Brain plasticity as a basis of the development of rehabilitation procedures for hemiplegia. *Scandinavian Journal Rehabilitation Medicine, 13,* 73–84.

Baddeley, A. D., & Hitch, G. (1974). Working memory. In G. H. Bower (Ed.), *The psychology of learning and motivation* (Vol. 8). New York: Academic Press.

Bailey, P., & von Bonin, G. (1951). *The isocortex of man.* Urbana: University of Illinois.

Basso, A., de Renzi, E., Faglioni, P., Scotti, G., & Spinnler, H. (1973). Neuropsychological evidence for the existence of cerebral areas critical to performance of intelligence tasks. *Brain, 96,* 715–728.

Basso, A., Faglioni, P., & Vignolo, L. A. (1975). Étude controlée de la reéducation du langage dans l'aphasie: Comparaison entre aphasiques traités et non-traité. *Revue Neurologique, 131,* 607–614.

Basso, A., Capitani, E., & Vignolo, L. A. (1979). Influence of rehabilitation on language skills in aphasic patients: A controlled study. *Archives of Neurology, 36*, 190–196.

Basso, A., Capitani, E., Luzzati, C., & Spinnler, H. (1981). Intelligence and left hemisphere disease. *Brain, 104*, 721–734.

Basso, A., Spinnler, H., Vallar, G., & Zanobio, M. E. (1982). Left hemisphere damage and selective impairment of auditory verbal short term memory. A case study. *Neuropsychologia, 20*, 263–274.

Bay, E. (1962). Aphasia and non-verbal disorders of language. *Brain, 3*, 412–426.

Bay, E. (1966). The classification of disorders of speech. *Cortex, 3*, 26–31.

Beauvois, M. F., & Lhermitte, F. (1975). Repetition immediate et apprentissage d'une série de mots chez 30 sujets aphasiques. *Neuropsychologia, 13*, 247–251.

Benson, D. F. (1977). The third alexia. *Archives of Neurology, 34*, 327–331.

Benson, D. F., & Geschwind, N. (1969). The alexias. In P. J. Vinken & G. W. Bruyn (Eds.), *Handbook of clinical neurology*. Amsterdam: North Holland Publishing Company.

Benson, D. F., & Geschwind, N. (1977). The aphasias and related disturbances. In A. B. Baker & L. H. Baker (Eds.), *Handbook of Clinical Neurology* (Vol. 1). New York: Harper and Row.

Benson, D. F., Sheremata, W. A., Bouchard, R., Segarra, J. M., Price, D., & Geschwind, N. (1973). Conduction aphasia: A clinicopathological study. *Archives of Neurology, 28*, 339–346.

Benton, A. L. (1961). The fiction of the "Gerstmann syndrome." *Journal of Neurology, Neurosurgery and Psychiatry, 24*, 176–181.

Benton, A. L. (1967). Problems of test construction in the field of aphasia. *Cortex, 3*, 32–53.

Black, F. W., & Strub, R. L. (1976). Constructional apraxia in patients with discrete missile wounds of the brain. *Cortex, 12*, 212–220.

Black, F. W., & Strub, R. L. (1978). Digit repetition performance in patients with focal brain damage. *Cortex, 14*, 12–21.

Blunk, R., DeBleser, R., Willmes, K., & Zeumer, H. (1981). A refined method to relate morphological and functional aspects of aphasia. *European Neurology, 20*, 69–79.

Bogen, J. E., & Bogen, G. M. (1976). Wernicke's region—Where is it? In S. R. Harnad *et al.* (Eds.), *Origins and evolution of language and speech*. New York: New York Academy of Sciences.

Borod, J. C., Carper, M., & Goodglass, H. (1982). WAIS performance IQ in aphasia as a function of auditory comprehension and constructional apraxia. *Cortex, 18*, 199–210.

Bradshaw, J. L., & Nettleton, N. C. (1981). The nature of hemispheric specialization in man. *The Behavioral and Brain Sciences, 4*, 51–92.

Broadbent, D. E. (1984). The maltese cross: A new simplistic model for memory. *The Behavioral and Brain Sciences, 7*, 55–94.

Broca, P. (1861). Remarques sur la siège de la faculté du langage articulé; suivies d'une observation d'aphemie (perte de la parole). *Bulletin de la Société Anatomique, 6,* 330–357.

Broman, T., Lindholm, A., & Melin, B. (1967). Rehabilitering av afasipasienter. *Läkartidningen, 64,* 4595–4600.

Brown, J. W. (1979). Language representation in the brain. In H. Steklis & M. Raleigh (Eds.), *Neurobiology of social communication in primates.* New York: Academic Press.

Brown, J. W., & Jaffe, J. (1975). Hypothesis on cerebral dominance. *Neuropsychologia, 13,* 107–110.

Brunner, R. J., Kornhuber, H. H., Seemuller, E., Suger, G., & Wallesch, C.-W. (1982). Basal ganglia participation in language pathology. *Brain and Language, 16,* 281–299.

Brust, J. C. M., Schafer, S., Richter, R., & Bruun, B. (1976). Aphasia in acute stroke. *Stroke, 7,* 167 171.

Bub, D., & Kertesz, A. (1982). Deep agraphia. *Brain and Language, 17,* 146–165.

Butfield, E., & Zangwill, O. L. (1946). Re-education in aphasia: A review of 70 cases. *Journal of Neurology, Neurosurgery and Psychiatry, 7,* 75–79.

Caplan, D. (1982). Representation in neurology and linguistics. In M. A. Arbib, D. Caplan, & J. C. Marshall (Eds.), *Neural models of language processes.* New York: Academic Press.

Caramazza, A., & Berndt, R. S. (1978). Semantic and syntactic processes in aphasia: A review of the literature. *Psychological Bulletin, 85,* 898–918.

Carson, D. H., Carson, F. E., & Tikofsky, R. S. (1968). On learning characteristics of the adult aphasic. *Cortex, 4,* 92–112.

Castro-Caldas, A., & Botelho, M. A. S. (1980). Dichotic listening in the recovery of aphasia after stroke. *Brain and Language, 10,* 145–151.

Cattell, R. B. (1978). *The scientific use of factor analysis in behavioural and life sciences.* New York: Plenum Press.

Cermak, L. S., & Moreines, J. (1976). Verbal retention deficits in aphasic and amnestic patients. *Brain and Language, 3,* 16–27.

Cermak, L. S., & Tarlow, S. (1978). Aphasic and amnesic patients' verbal vs. nonverbal retentive abilities. *Cortex, 14,* 32–40.

Chomsky, N. (1965). *Aspects of the theory of syntax.* Cambridge, Mass.: MIT Press.

Christensen, A.-L. (1975). *Luria's neuropsychological investigation.* Copenhagen: Munksgaard.

Cicone, M., Wapner, W., Foldi, N., Zurif, E., & Gardner, H. (1979). The relation between gesture and language in aphasic communication. *Brain and Language, 8,* 324–349.

Coltheart, M., Patterson, K., & Marshall, J. C. (Eds.). (1980). *Deep dyslexia.* London: Routledge & Kegan Paul.

Conrad, R. (1964). Acoustic confusion in immediate memory. *British Journal of Psychology, 55,* 75–84.

Corsi, P. M. (1972). *Human memory and the medial temporal region of the brain.* Ph.D. dissertation, McGill University, Montreal.

Costa, L. D. (1975). The relation of visuospatial dysfunction to digit span performance in patients with cerebral lesions. *Cortex, 11,* 31–36.

Costa, L. D. (1976). Interset variability on the Raven Coloured Progressive Matrices as an indicator of specific ability deficit in brain-lesioned patients. *Cortex, 12,* 31–40.

Crowder, R. G. (1976). *Principles of learning and memory.* Hillsdale, N.J.: Lawrence Erlbaum.

Crowder, R. G., & Morton, J. (1969). Precategorical acoustic storage (PAS). *Perception and Psychophysics, 5,* 365–373.

Culton, G. L. (1969). Spontaneous recovery from aphasia. *Journal of Speech and Hearing Research, 12,* 825–832.

Damasio, A. R., Damasio, H., Rizzo, M., Varney, N., & Gersh, F. (1982). Aphasia with nonhemorrhagic lesions in the basal ganglion and internal capsule. *Archives of Neurology, 39,* 15–20.

Déjerine, J. (1892). Contribution a l'étude anatomo-pathologique et clinique des différentes variétés de cécité verbal. *Mémoires de la Société de Biologie, 4,* 61–90.

Déjerine, J. (1914). *Sémiéologie des affections du système nerveux.* Paris: Masson.

Denenberg, V. H. (1978). Dilemmas and designs for developmental research. In C. L. Ludlow & E. Doran-Quine (Eds.), *The neurological bases of language disorders in children: Methods and directions for research.* Bethesda, Maryland: U.S. Department of Health, Education and Welfare.

Denenberg, V. H. (1981). Hemispheric laterality in animals and the effects of early experience. *The Behavioral and Brain Sciences, 4,* 1–50.

Denes, F., Semenza, C., Stoppa, E., & Gradenigo, G. (1978). Selective improvement by unilateral brain-damaged patients on Raven Coloured Progressive Matrices. *Neuropsychologia, 16,* 749–752.

Dennis, M. (1976). Dissociated naming and locating of body parts after left anterior temporal lobe resection: An experimental case study. *Brain and Language, 3,* 147–163.

Dennis, M., & Whitaker, H. A. (1977). Hemispheric equipotentiality and language acquisition. In S. J. Segalowitz & F. A. Gruber (Eds.), *Language development and neurological theory.* New York: Academic Press.

de Renzi, E. (1982). *Disorders of space exploration and cognition.* New York: Wiley.

de Renzi, E., & Faglioni, P. (1965). The comparative effect of intelligence and vigilance tests in detecting hemispheric cerebral damage. *Cortex, 1,* 410–433.

de Renzi, E., Faglioni, P., & Previdi, P. (1977). Spatial memory and hemispheric locus of lesion. *Cortex, 13,* 424–433.

de Renzi, E., Faglioni, P., & Villa, P. (1977). Sequential memory for figures in brain-damaged patients. *Neuropsychologia, 15,* 42–50.

de Renzi, E., & Nichelli, P. (1975). Verbal and non-verbal short term memory impairment following hemispheric damage. *Cortex, 11,* 341–354.

de Renzi, E., & Spinnler, H. (1967). Impaired performance on color tasks in patients with hemispheric damage. *Cortex, 3,* 194–216.

Duffy, R. J., & Duffy, J. R. (1981). Three studies of deficit in pantomime expression and pantomime recognition in aphasia. *Journal of Speech and Hearing Research, 24,* 70–84.

Duffy, R. J., & Liles, B. Z. (1979). A translation of Finkelnburg's (1870) lecture on aphasia as "asymbolia" with commentary. *Journal of Speech and Hearing Disorders, 44,* 156–168.

Ebenholz, S. M. (1972). Serial learning and dimensional organization. In G. H. Bower (Ed.), *The psychology of learning and motivation* (Vol. 5). New York: Academic Press.

Efron, R. (1963). Temporal perception, aphasia and déjà vu. *Brain, 86,* 403–424.

Engvik, H., Hjerkinn, O., & Seim, S. (1980). *WAIS: Norsk utgave.* Oslo: Norsk Psykologforening.

Exner, S. (1881). *Untersuchungen über die Localisation der Functionen in der Grosshirnrinde der Menschen.* Wien: Braunmueller.

Fedio, P., & van Buren, J. M. (1975). Memory and perceptual deficits during electrical stimulation in the left and right thalamus and parietal subcortex. *Brain and Language, 2,* 78–100.

Finger, S. (1978). Environmental attenuation of brain-lesion symptoms. In S. Finger (Ed.), *Recovery from brain damage.* New York: Plenum Press.

Fodor, J., & Bever, T. G. (1965). The psychological reality of linguistic segments. *Journal of Verbal Learning and Verbal Behaviour, 4,* 414–420.

Fredriksen, A. L., & Lernæs, L. K. (1984). *Sex differences in cerebral organization of language.* (In Norwegian). Unpublished thesis, Department of Psychology, University of Oslo.

Friederici, A. D., Schoenle, P. W., & Goodglass, H. (1981). Mechanisms underlying writing and speech in aphasia. *Brain and Language, 13,* 212–222.

Friedman, A., & Polson, M. C. (1981). Hemispheres as independent resource systems: Limited-capacity processing and cerebral specialization. *Journal of Experimental Psychology: Human Perception and Performance, 7,* 1031–1058.

Frostig, M. (1966). *Developmental test of visual perception.* Palo Alto, Calif.: Consulting Psychologists Press.

Galaburda, A. M. (1982). Histology, architectonics, and asymmetry of language areas. In M. A. Arbib, D. Caplan, & J. C. Marshall (Eds.), *Neural models of language processes.* New York: Academic Press.

Galaburda, A. M., LeMay, M., Kemper, T. L., & Geschwind, N. (1978). Right-left asymmetries in the brain. *Science, 199,* 852–856.

Gardner, H., Zurif, E., Berry, T., & Baker, E. (1976). Visual communication therapy in aphasia. *Neuropsychologia, 14,* 275–299.

Gazzaniga, M., & Ledoux, J. (1978). *The integrated mind.* New York: Plenum Press.

Gazzaniga, M. S., Velletri Glass, A., Sarno, M. T., & Posner, J. B. (1973).

Pure word deafness and hemispheric dynamics: A case history. *Cortex,*
9, 136–143.

Geschwind, N. (1965). Disconnection syndromes in animals and man. *Brain,*
88, 234–294, 585–644.

Geschwind, N. (1967a). The apraxias. In E. W. Strauss & R. M. Griffith
(Eds.), *Phenomenology of will and action.* Pittsburgh: Duquesne University
Press.

Geschwind, N. (1967b). The variety of naming errors. *Cortex, 3,* 97–112.

Geschwind, N. (1979). Specializations of the human brain. *Scientific American,*
241, 180–199.

Geschwind, N., & Fusillo, M. (1966). Color-naming defects in association
with alexia. *Archives of Neurology, 15,* 137–146.

Geschwind, N., & Levitsky, W. (1968). Human brain: Left-right asymmetry
in temporal speech region. *Science, 161,* 186–187.

Geschwind, N., Quadfasel, F., & Segarra, J. M. (1968). Isolation of the speech
area. *Neuropsychologia, 6,* 327–340.

Glanzer, M. (1972). Storage mechanisms in recall. In G. H. Bower (Ed.), *The
psychology of learning and motivation* (Vol. 5). New York: Academic Press.

Glick, S. D., & Zimmerberg, B. (1978). Pharmacological modification of brain
lesion syndromes. In S. Finger (Ed.), *Recovery from brain damage: Research
and theory.* New York: Plenum Press.

Goldstein, K. (1948). *Language and language disturbances.* New York: Grune
and Stratton.

Goodglass, H., Denes, G., & Calderon, M. (1974). The absence of covert
verbal mediation in aphasia. *Cortex, 10,* 264–269.

Goodglass, H., Gleason, J. B., & Hyde, M. (1970). Some dimensions of auditory language comprehension in aphasia. *Journal of Speech and Hearing
Research, 13,* 595–606.

Goodglass, H., & Kaplan, E. (1963). Disturbance of gesture and pantomime
in aphasia. *Brain, 86,* 703–720.

Goodglass, H., & Kaplan, E. (1972). *The assessment of aphasia.* Philadelphia:
Lea and Febiger.

Goodglass, H., Klein, B., Carey, P. W., & Jones, K. J. (1966). Specific semantic
word categories in aphasia. *Cortex, 2,* 74–89.

Gordon, W. P. (1983). Memory disorders in aphasia-I. Auditory immediate
recall. *Neuropsychologia, 21,* 325–339.

Green, E., & Howes, D. (1977). The nature of conduction aphasia. In H.
Whitaker & H. A. Whitaker (Eds.), *Studies in neurolinguistics* (Vol. 3).
New York: Academic Press.

Hamsher, K. (1982). Intelligence and aphasia. In M. T. Sarno (Ed.), *Acquired
aphasia.* New York: Academic Press.

Hamsher, K., Benton, A. L., & Digre, K. (1980). Serial digit learning: Normative and clinical aspects. *Journal of Clinical Neuropsychology, 2,* 39–50.

Hasher, L., & Zacks, R. T. (1979). Automatic and effortful processes in memory. *Journal of Experimental Psychology: General, 108,* 356–388.

Head, H. (1915). Hughling Jackson on aphasia and kindred affections of speech. *Brain, 38*, 1–27.

Head, H. (1926). *Aphasia and kindred disorders of speech.* Cambridge: Cambridge University Press.

Hécaen, H., & Assal, G. A. (1970). Comparison of constructive deficits following right and left hemispheric lesions. *Neuropsychologia, 8*, 289–303.

Hécaen, H., & Consoli, S. (1973). Analyse des troubles du langage au cours des lésions de l'aire de Broca. *Neuropsychologia, 11*, 371–388.

Hécaen, H., & Dubois, J. (1971). La neurolinguistique. In G. E. Perren & J. L. M. Trim (Eds.), *Applications of linguistics.* Cambridge: Cambridge University Press, 1971.

Hécaen, H., & Kremin, H. (1976). Neurolinguistic research on reading disorders resulting from left hemisphere lesions: Aphasic and "pure" alexia. In H. Whitaker & H. A. Whitaker (Eds.), *Studies in neurolinguistics* (Vol. 2). New York: Academic Press.

Heilman, K. M., Safran, A., & Geschwind, N. (1971). Closed head trauma and aphasia. *Journal of Neurology, Neurosurgery and Psychiatry, 34*, 265–269.

Heilman, K. M., Scholes, R., & Watson, R. T. (1976). Defects of immediate memory in Broca's and conduction aphasia. *Brain and Language, 3*, 201–208.

Henschen, S. E. (1922). *Klinische und anatomische Beitraege zur Pathologie des Gehirns* (Vols. 5–7). Stockholm: Nordiska Bokhandel.

Hirst, W. (1982). The amnesic syndrome: Descriptions and explanations. *Psychological Bulletin, 91*, 435–460.

Howes, D., & Geschwind, N. (1964). Quantitative studies of aphasic language. In D. Rioch & E. Weinstein (Eds.), *Disorders of communication.* Baltimore: Williams and Wilkins.

Huber, W., Stachowiack, F.-J., Poeck, K., & Kerschensteiner, M. (1975). Die Wernicke Aphasie. *Journal of Neurology, 210*, 77–97.

Huber, W., Poeck, K., Weniger, D., & Willmes, K. (1983). *Der Aachener Aphasie Test (AAT).* Göttingen: Hogrefe.

Jackson, J. H. (1878). On affections of speech from disease of the brain. *Brain, 1*, 304–330.

Jakobson, R. (1971). Aphasia as a linguistic topic. Reprinted in Roman Jakobson, *Selected Writings* (Vol. 2). The Hague: Mouton.

Johnson, J. P., Sommers, R. K., & Weidener, W. E. (1977). Dichotic ear preference in aphasia. *Journal of Speech and Hearing Research, 20*, 116–129.

Jones, E. G., & Powell, T. P. S. (1970). An anatomical study of converging sensory pathways within the cerebral cortex of the monkey. *Brain, 93*, 793–820.

Jones, L. V., & Wepman, J. M. (1961). Dimensions of language performance in aphasia. *Journal of Speech and Hearing Research, 4*, 220–232.

Kenin, M., & Swisher, L. (1972). A study of pattern of recovery in aphasia. *Cortex, 8*, 56–68.

Kerschensteiner, M., Poeck, K., & Brunner, E. (1972). The fluency-nonfluency dimension in the classification of aphasic speech. *Cortex, 8,* 233–247.

Kerschensteiner, M., Poeck, K., Huber, W., Stachowiack, F.-J., & Weniger, D. (1975). Die Broca Aphasie. *Journal of Neurology, 217,* 223–242.

Kertesz, A. (1979). *Aphasia and associated disorders.* New York: Grune and Stratton.

Kertesz, A., & Benson, D. F. (1970). Neologistic jargon: A clinico-pathological study. *Cortex, 6,* 362–386.

Kertesz, A., & Hooper, P. (1982). Praxis and language: The extent and variety of apraxia in aphasia. *Neuropsychologia, 20,* 275–286.

Kertesz, A., & McCabe, P. (1975). Intelligence and aphasia: Performance of aphasics on Raven's Coloured Progressive Matrices (RCPM). *Brain and Language, 2,* 387–395.

Kertesz, A., & McCabe, P. (1977). Recovery patterns and prognosis in aphasia. *Brain, 100,* 1–18.

Kertesz, A., & Phipps, J. B. (1977). Numerical taxonomy of aphasia. *Brain and Language, 4,* 1–10.

Kertesz, A., & Phipps, J. (1980). The numerical taxonomy of acute and chronic aphasic syndromes. *Psychological Research, 41,* 179–198.

Kertesz, A., & Poole, E. (1974). The aphasia quotient: The taxonomic approach to measurement of aphasic disability. *Canadian Journal of Neurological Sciences, 1,* 7–16.

Kertesz, A., & Sheppard, A. (1981). The epidemiology of aphasic and cognitive impairment in stroke. *Brain, 104,* 117–128.

Kertesz, A., Harlock, W., & Coates, R. (1979). Computer tomographic localization, lesion size, and prognosis in aphasia and nonverbal impairment. *Brain and Language, 8,* 34–51.

Kertesz, A., Lesk, D., & McCabe, P. (1977). Isotope localization of infarcts in aphasia. *Archives of Neurology, 34,* 590–601.

Kimura, D. (1979). Neuromotor mechanisms in the evolution of human communication. In H. Steklis & M. Raleigh (Eds.), *Neurobiology of social communication in primates.* New York: Academic Press.

Kimura, D. (1983). Sex differences in cerebral organization for speech and praxis functions. *Canadian Journal of Psychology, 37,* 19–35.

Kinsbourne, M. (1971). The minor cerebral hemisphere as a source of aphasic speech. *Archives of Neurology, 25,* 302–306.

Kinsbourne, M. (1982). Hemispheric specialization and the growth of human understanding. *American Psychologist, 37,* 411–420.

Kinsbourne, M., & Hicks, R. E. (1978). Functional cerebral space: A model for overflow, transfer and interference effects in human performance: A tutorial review. In J. Requin (Ed.), *Attention and performance* (Vol. 7). Hillsdale, N.J.: Lawrence Erlbaum.

Kinsbourne, M., & Warrington, E. K. (1964). Observations on colour agnosia. *Journal of Neurology, Neurosurgery and Psychiatry, 27,* 296–299.

Kohn, B., & Dennis, M. (1974). Selective impairments of visuospatial abilities in infantile hemiplegics after right hemidecortication. *Neuropsychologia*, *12*, 505–512.

Kolers, P. A., Palef, S. R., & Stelmach, L. B. (1980). Graphemic analysis underlying literacy. *Memory and Cognition, 8,* 322–328.

Kreindler, A., & Fradis, A. (1968). *Performances in aphasia.* Paris: Gauthier-Villars.

Laine, T., & Marttila, R. J. (1981). Pure agraphia: A case study. *Neuropsychologia, 19,* 311–316.

Lamendella, J. T. (1977). The limbic system in human communication. In H. Whitaker & H. A. Whitaker (Eds.), *Studies in neurolinguistics* (Vol. 3). New York: Academic Press.

Lassen, N. A., Ingvar, D. H., & Skinhøj, E. (1978). Brain function and blood flow. *Scientific American, 239,* 50–59.

Lebrun, Y., & Hoops, R. (Eds.). (1974). *Intelligence and aphasia.* Amsterdam: Swets and Zeitlinger.

Lecours, A. R., & Rouillon, F. (1976). Neurolinguistic analysis of jargon aphasia and jargon agraphia. In H. Whitaker & H. A. Whitaker (Eds.), *Studies in neurolinguistics* (Vol. 2). New York: Academic Press.

Lehman, E. B. (1982). Memory for modality: Evidence for an automatic process. *Memory and Cognition, 10,* 554–564.

Lehmkuhl, G., Poeck, K., & Willmes, K. (1983). Ideomotor apraxia and aphasia: An examination of types and manifestations of apraxic symptoms. *Neuropsychologia, 21,* 199–212.

Leischner, A. (1972). Über den Verlauf und die Einteilung der aphasischen Syndrome. *Archiv für Psychiatrie und Nervenkrankheit, 216,* 219–231.

Lenneberg, E. H. (1967). *Biological foundations of language.* New York: Wiley.

Levine, D., & Calvanio, R. (1982). The neurology of reading disorders. In M. A. Arbib, D. Caplan, & J. C. Marshall (Eds.), *Neural models of language processes.* New York: Academic Press.

Levy, B. A. (1983). Proofreading familiar text: Constraints on visual processing. *Memory and Cognition, 11,* 1–12.

Lichtheim, L. (1885). On aphasia. *Brain, 7,* 443–484.

Liepmann, H. (1900). Das Krankheitsbild der Apraxie ("motorischen Asymbolie"). *Monatschrift für Psychologie und Neurologie, Vol. 8.*

Liepmann, H. (1915). Diseases of the brain. In C. W. Barr (Ed.), *Curschmanns textbook on nervous diseases* (Vol. 1). Philadelphia: Blakiston.

Lomas, J., & Kertesz, A. (1978). Patterns of spontaneous recovery in aphasic groups: A study of adult stroke patients. *Brain and Language, 5,* 388–401.

Luria, A. R. (1966). *Higher cortical function in man.* New York: Basic Books.

Luria, A. R. (1970). *Traumatic aphasia* (English ed.). The Hague: Mouton.

Luria, A. R. (1973). *The working brain.* London: Penguin.

Luria, A. R. (1977). On quasi-aphasic speech disturbances in lesions of the deep structures of the brain. *Brain and Language, 4,* 432–459.

Marie, P. (1906). Révision de la question de l'aphasie: La troisième circonvolution frontale gauche ne joue aucun role spécial dans la fonction du langage. *Semaine Medicale, 26,* 241–247.

Marquardsen, J. (1969). The natural history of acute cerebrovascular disease. *Acta Neurologica Scandinavica, 45*, Supp. 38.

Marshall, J. C. (1982). What is a symptom-complex? In M. A. Arbib, D. Caplan, & J. C. Marshall (Eds.), *Neural models of language processes.* New York: Academic Press.

Marshall, J. C., & Newcombe, F. (1973). Patterns of paralexia: A psycholinguistic approach. *Psycholinguistic Research, 2,* 175–199.

Matsui, T., & Hirano, A. (1978). *An atlas of the human brain for computerized tomography.* New York: Fischer.

McFarling, D., Rothi, L. J., & Heilman, K. M. (1982). Transcortical aphasia from ischemic infarcts of the thalamus: A report on two cases. *Journal of Neurology, Neurosurgery and Psychiatry, 45,* 107–112.

McFie, J. (1975). *Assessment of organic intellectual impairment.* New York: Academic Press.

McGlone, J. (1980). Sex differences in human cerebral asymmetry: A critical survey. *The Behavioral and Brain Sciences, 3,* 215–263.

Milner, B. (1962). Laterality effects in audition. In V. B. Mountcastle (Ed.), *Interhemispheric relations and cerebral dominance.* Baltimore: The Johns Hopkins Press.

Milner, B. (1974). Hemispheric specialization: Scope and limits. In F. O. Schmitt & F. G. Worden (Eds.), *The neurosciences: Third study program.* Cambridge, Mass: MIT Press.

Milner, B., Branch, C., & Rasmussen, T. (1964). Observations on cerebral dominance. In A. V. S. de Reuch & M. O'Connor (Eds.), *Disorders of language.* Boston: Little, Brown.

Mohr, J. P. (1976). Broca's area and Broca's aphasia. In H. Whitaker & H. A. Whitaker (Eds.), *Studies in Neurolinguistics* (Vol. 1). New York: Academic Press.

Mohr, J. P., Pessin, M., Finkelstein, S., Finkelstein, H., Duncan, G., & Grand Davis, K. (1978). Broca aphasia: Pathological and clinical aspects. *Neurology, 28,* 311–324.

Mohr, J. P., Sidman, M., Stoddard, L. T., Leicester, J., & Rosenberger, P. B. (1973). Evolution of the deficit in total aphasia. *Neurology, 23,* 1302–1312.

Morton, J. (1979). Facilitation in word recognition: Experiments causing change in the logogen model. In P. A. Kolers, M. Wrolstad, & H. Bouma (Eds.), *Processing of visible language* (Vol. 1). New York: Plenum Press.

Moscowitch, M. (1977). The development of lateralization of language functions and its relation to cognitive and linguistic development: A review and some theoretical speculations. In S. J. Segalowitz & F. A. Gruber (Eds.), *Language development and neurological theory.* New York: Academic Press.

Mountcastle, V. B. (1957). Modality and topographic properties of single neurons of cat's somatic sensory cortex. *Journal of Neurophysiology, 20,* 408–434.

Naeser, A., & Hayward, R. W. (1978). Lesion localization in aphasia with cranial computed tomography and the Boston Diagnostic Aphasia Exam. *Neurology, 28*, 545–551.

Navon, D., & Gopher, D. (1979). On the economy of the human-processing system. *Psychological Review, 86*, 214–255.

Newcombe, F. (1969). *Missile wounds of the brain.* London: Oxford University Press.

Nie, N. H., Hadlai Hull, C., Jenkins, G. J., Steinbrenner, K., & Brent, D. H. (1975). *SPSS: Statistical Package for the Social Sciences.* New York: McGraw-Hill.

Nunally, J. C. (1967). *Psychometric theory.* New York: McGraw-Hill.

Ohler, I. K., Albert, M. L., Goodglass, H., & Benson, D. F. (1978). Aphasia type and aging. *Brain and Language, 6*, 318–322.

Orgass, B., Hartje, W., Kerschensteiner, M., & Poeck, K. (1972). Aphasie und nicht-sprachliche Intelligenz. *Nervenarzt, 43*, 623–627.

Orgass, B., Poeck, K., & Kerschensteiner, M. (1974). Das Verstandnis für Nomina mit spezifischer Referenz bei aphasischen Patienten. *Zeitschrift für Neurologie, 206*, 95–102.

Oxbury, J. M., Oxbury, S. M., & Humphrey, N. K. (1969). Variety of colour anomia. *Brain, 92*, 847–860.

Pandya, D. N., & Galaburda, A. M. (1980). Role of architectonics and connections in the study of primate brain evolution. *American Journal of Physical Anthropology, 52*, 197.

Penfield, W., & Roberts, L. (1959). *Speech and brain mechanisms.* Princeton, N.J.: Princeton University Press.

Petlund, C.-F. (1970). *Prevalence and invalidity from stroke in Aust-Agder county of Norway.* Oslo: Universitetsforlaget.

Pettit, J. M., & Noll, J. D. (1979). Cerebral dominance in aphasia recovery. *Brain and Language, 7*, 191–200.

Poeck, K. (1983a). Ideational apraxia. *Journal of Neurology, 230*, 1–5.

Poeck, K. (1983b). What do we mean by "aphasic syndromes"? A neurologist's view. *Brain and Language, 20*, 79–89.

Poeck, K., & Huber, W. (1977). To what extent is language a sequential activity? *Neuropsychologia, 15*, 359–363.

Poeck, K., & Lehmkuhl, G. (1980). Das Syndrom der ideatorischen Apraxie und seine Lokalisation. *Nervenartzt, 51*, 217–225.

Poeck, K., & Stachowiak, F.-J. (1975). Farbbenennungsstoerungen bei aphasischen und nichtaphasischen Hirnkranken. *Journal of Neurology, 209*, 95–102.

Pribram, K. H. (1971). *Languages of the brain.* Englewood Cliffs, N.J.: Prentice-Hall.

Prins, R., Snow, C., & Wagenaar, E. (1978). Recovery from aphasia: Spontaneous speech versus language comprehension. *Brain and Language, 5*, 192–211.

Raisman, G., & Field, P. M. (1973). A quantitative investigation of the development of collateral reinnervation of the septal nuclei. *Brain Research, 50,* 241–264.

Rasmussen, T., & Milner, B. (1975). Clinical and surgical studies of the cerebral speech areas in man. In K. S. Zülch *et al.* (Eds.), *Cerebral localization.* New York: Springer.

Ratliffe, F. (1961). Inhibitory interaction and the detection and enhancement of contours. In S. W. Rosenblith (Ed.), *Sensory communication.* Cambridge, Mass.: MIT Press.

Raven, J. C. (1960). *Coloured Progressive Matrices.* London: H. K. Lewis.

Razy, A., Janotta, F. S., & Lehner, L. H. (1969). Aphasia resulting from occlusion of the left anterior cerebral artery. *Archives of Neurology, 36,* 221–225.

Reinvang, I. (1981). *Patterns of restitution in aphasia.* Paper presented at INS European conference, Bergen, Norway, June.

Reinvang, I. (1983). *The systemically organized neural basis of language: Aphasia syndromes and their recovery.* Doctoral dissertation, University of Oslo.

Reinvang, I., & Dugstad, G. (1981). Aphasia typology and lesion localization with computed axial tomography. *Scandinavian Journal of Rehabilitation Medicine, 13,* 85–88.

Reinvang, I., & Engvik, H. (1980a). Language recovery in aphasics from 3 to 6 months after stroke. In M. T. Sarno & O. Höök (Eds.), *Aphasia: Assessment and treatment.* Stockholm: Almquist & Wiksell.

Reinvang, I., & Engvik, H. (1980b). *Norsk Grunntest for Afasi.* Oslo: Universitetsforlaget. (Danish edition, Copenhagen: Dansk Psykologisk Forlag, 1984.)

Reitan, R. M., & Davison, L. A. (Eds.). (1974). *Clinical neuropsychology: Current status and applications.* Washington, D.C.: Winston.

Richardson, J. T. E. (1977). Functional relationship between forward and backward digit repetition and a non-verbal analogue. *Cortex, 13,* 317–320.

Robinson, B. W. (1976). Limbic influences on human speech. In S. Harnad *et al.* (Eds.), *Origins and evolution of language and speech.* New York: The New York Academy of Sciences.

Robinson, R. G., & Benson, D. F. (1981). Depression in aphasic patients: Frequency, severity and clinico-pathological correlations. *Brain and Language, 14,* 282–291.

Rothi, L. J., & Hutchinson, E. C. (1981). Retention of verbal information by rehearsal in relation to the fluency of verbal output in aphasia. *Brain and Language, 12,* 347–359.

Royce, J., & Powell, A. (1983). *Theory of personality and individual differences: Factors, systems, and processes.* Englewood Cliffs, N.J.: Prentice Hall.

Rubens, A. B. (1976). Transcortical motor aphasia. In H. Whitaker & H. A. Whitaker (Eds.), *Studies in neurolinguistics* (Vol. 1). New York: Academic Press.

Rudel, R. G., & Denckla, M. B. (1974). Relation of forward and backward digit repetition to neurological impairment in children with learning disabilities. *Neuropsychologia, 12,* 109–118.

Russo, M., & Vignolo, L. A. (1967). Visual figure-ground discrimination in patients with unilateral cerebral disease. *Cortex, 3,* 113–127.

Sarno, M. T. (1976). The status of research in recovery from aphasia. In Y. Lebrun & R. Hoops (Eds.), *Recovery in aphasics.* Amsterdam: Swets and Zeitlinger.

Sarno, M. T., & Levita, E. (1971). Natural course of recovery in severe aphasia. *Archives of Physical Medicine and Rehabilitation, 52,* 175–179.

Sarno, M. T., & Levita, E. (1979). Recovery in treated aphasia in the first year post-stroke. *Stroke, 10,* 663–670.

Sarno, M. T., & Levita, E. (1981). Some observations on the nature of recovery in global aphasia. *Brain and Language, 13,* 1–12.

Sarno, M. T., Silverman, M., & Sands, E. (1970). Speech therapy and language recovery in severe aphasia. *Journal of Speech and Hearing Research, 13,* 607–623.

Schmachtemberg, A., Hundgen, R., & Zeumer, H. (1983). Ein EDV-adaptiertes Rastermodell des Gehirns zur topographischen Analyse von Läsionen im kranialen Computertomogramm. *Fortschritte Röntgenstrahlung, 139,* 499–502.

Schneider, G. E. (1973). Early lesions of superior colliculus: Factors affecting the formation of abnormal retinal projections. *Brain Behavior and Evolution, 8,* 73–109.

Schuell, H. (1974). *Aphasia theory and therapy.* New York: Macmillan.

Schuell, H., Jenkins, J. J., & Carroll, J. B. (1962). A factor analysis of the Minnesota test for differential diagnosis of aphasia. *Journal of Speech and Hearing Research, 5,* 349–369.

Schuell, H., Jenkins, J. J., & Jimenez-Pabon, E. (1965). *Aphasia in adults.* New York: Harper and Row.

Schwartz, M. F. (1984). What the classical aphasia categories can't do for us and why. *Brain and Language, 21,* 3–8.

Selnes, O. A. (1974). The corpus callosum: Some anatomical and functional considerations with special reference to language. *Brain and Language, 1,* 111–140.

Shallice, T. (1979). Case study approach in neuropsychological research. *Journal of Clinical Neuropsychology, 1,* 183–211.

Shallice, T., & Warrington, E. K. (1974). The dissociation between short term retention of meaningful sounds and verbal material. *Neuropsychologia, 12,* 553–555.

Smith, A., Champoux, R., Levi, J., London, R., & Muraski, A. (1972). *Diagnosis, intelligence and rehabilitation of chronic aphasics.* Ann Arbor: University of Michigan.

Snedecor, G. W., & Cochran, W. G. (1967). *Statistical methods.* Ames: The Iowa University Press.

Spelke, E., Hirst, W., & Neisser, U. (1976). Skills of divided attention. *Cognition, 4*, 215–230.

Sperling, G. (1960). The information available in brief visual presentations. *Psychological Monographs, 74*, No. 498.

Stachowiak, F. J., Huber, W., Kerschensteiner, M., Poeck, K., & Weniger, D. (1977). Die globale Aphasie. *Journal of Neurology, 214*, 75–87.

Studdert-Kennedy, M., & Shankweiler, D. P. (1970). Hemispheric specialization for speech perception. *Journal of the Acoustical Society of America, 48*, 579–594.

Subirana, A. (1969). Handedness and cerebral dominance. In P. Vinken & G. Bruyn (Eds.), *Handbook of clinical neurology* (Vol. 4). New York: Elsevier.

Sundet, K., & Engvik, H. (1984). *The validity of aphasic subtypes.* Paper presented at INS-European conference, Aachen, West Germany, June.

Swisher, L. P., & Hirsh, I. J. (1972). Brain damage and the ordering of two temporally successive stimuli. *Neuropsychologia, 10*, 137–152.

Szentagothai, J., & Arbib, M. A. (1975). *Conceptual models of neural organization.* Cambridge, Mass.: MIT Press.

Teuber, H. L. (1976). Complex functions of basal ganglia. In M. D. Yahr (Ed.), *The basal ganglia.* New York: Raven Press.

Teuber, H. L., & Weinstein, S. (1956). Ability to discover hidden figures after cerebral lesions. *Archives of Neurology and Psychiatry, 76*, 369–379.

Tulving, E. (1972). Episodic and semantic memory. In E. Tulving and W. Donaldson (Eds.), *Organization of memory.* New York: Academic Press.

Tulving, E. (1982). *Elements of episodic memory.* New York: Oxford University Press.

Tzortzis, C., & Albert, M. L. (1974). Impairment of memory for sequences in conduction aphasia. *Neuropsychologia, 12*, 355–366.

Ulrich, G. (1978). Interhemispheric functional relationship in auditory agnosia. *Brain and Language, 5*, 286–300.

Vanderplas, J. M., & Garvin, E. A. (1959). The association value of random shapes. *Journal of Experimental Psychology, 37*, 147–163.

Varney, N. R. (1978). Linguistic correlates of pantomime recognition in aphasic patients. *Journal of Neurology, Neurosurgery and Psychiatry, 41*, 564–568.

Vignolo, L. A (1964). Evolution of aphasia and language rehabilitation: A retrospective exploratory study. *Cortex, 1*, 344–367.

von Bertallanfy, L. (1948). *General systems theory.* New York: Braziller.

Wada, J., Clark, R., & Hamm, A. (1975). Cerebral hemispheric asymmetry in humans. *Archives of Neurology, 32*, 239–246.

Wallesch, C.-W., Kornhuber, H. H., Brunner, R. J., Kunz, T., Hollerbach, B., & Suger, G. (1983). Lesions of the basal ganglia, thalamus and deep white matter: Differential effects on language functions. *Brain and Language, 20*, 286–304.

Walley, R. E., & Weiden, T. D. (1973). Lateral inhibition and cognitive masking: A neuropsychological theory of attention. *Psychological Review, 80*, 284–302.

Warrington, E. K., & Shallice, T. (1969). The selective impairment of auditory verbal short-term memory. *Brain, 92*, 885–896.

Warrington, E. K., Logue, V., & Pratt, R. T. C. (1971). The anatomical localization of selective impairment of auditory verbal short-term memory. *Neuropsychologia, 9*, 377–387.

Wechsler, D. (1945). Standardized memory scale for clinical use. *Journal of Psychology, 19*, 87–95.

Wechsler, D. (1958). *The measurement and appraisal of adult intelligence.* Baltimore: Williams and Wilkins.

Weinberg, J., Diller, L., Gordon, W. A., Gerstman, L. J., Lieberman, A., Lakin, P., Hodges, G., & Ezrachi, O. (1977). Visual scanning training: Effect on reading related tasks in acquired right brain damage. *Archives of Physical Medicine and Rehabilitation, 58*, 479–486.

Weinstein, S. (1964). Deficits concomitant with aphasia or lesions of either hemisphere. *Cortex, 1*, 154–169.

Weisenburg, T., and McBride, K. (1935). *Aphasia.* New York: The Commonwealth Fund.

Weiss, P. (1969). The living system: determinism stratified. In A. Koestler and J. R. Smythies (Eds.), *Beyond reductionism.* Boston: Beacon.

Wepman, J. M. (1953). *Recovery from aphasia.* New York: Ronald.

Wernicke, C. (1874). *Der aphasische Symptomen-complex.* Breslau: Cohn und Weigert. Reprinted in *Boston Studies on the Philosophy of Science* (Vol. 4). Dordrecht: Reidl, 1964.

Whitaker, H. (1971). Neurolinguistics. In W. O. Dingwall (Ed.), *A survey of linguistic science.* College Park: University of Maryland.

White, N., & Kinsbourne, M. (1980). Does speech output control lateralize over time? Evidence from verbal-manual time-sharing tasks. *Brain and Language, 10*, 215–223.

Wiegel-Crump, C. A. (1976). Agrammatism and aphasia. In Y. Lebrun & R. Hoops (Eds.), *Recovery in aphasics.* Amsterdam: Swets and Zeitlinger.

Wiegel-Crump, C. A., & Koenigsknecht, R. (1973). Tapping the lexical store of the adult aphasic: Analysis of the improvement made in word retrieval skills. *Cortex, 9*, 410–418.

Willmes, K., Poeck, K., Weiniger, D., & Huber, W. (1980). Der Aachener Aphasie Test: Differentielle Validität. *Nervenartzt, 51*, 553–560.

Witkin, H. A., Goodenough, D. R., & Oltmann, P. K. (1977). *Psychological differentiation: Current status.* Princeton, N.J.: Educational Testing Service.

Wood, C. C. (1978). Variations on a theme by Lashley: Lesion experiments on the neural model of Anderson, Silverstein, Ritz and Jones. *Psychological Review, 85*, 582–591.

Woods, B. T., & Teuber, H. L. (1973). Early onset of complementary specialization of cerebral hemispheres in man. *Transactions of the American Neurological Association, 98*, 113–115.

Woolsey, C. N. (1958). Organization of somatic sensory and motor areas of the cerebral cortex. In H. F. Harlow & C. N. Woolsey (Eds.), *Biological and biochemical bases of behavior.* Madison: University of Wisconsin Press.

Yakovlev, P. I., & Lecours, A. R. (1967). The myelogenic cycles of regional maturation in the brain. In A. Minkowski (Ed.), *Regional development of the brain in early life.* Oxford: Blackwell.

Zangwill, O. (1964). Intelligence in aphasia. In A. V. S. de Reuch & M. O'Connor (Eds.), *Disorders of language.* Boston: Little, Brown.

Zurif, E. B., & Caramazza, A. (1976). Psycholinguistic structures in aphasia: Studies in syntax and semantics. In H. Whitaker & H. A. Whitaker (Eds.), *Studies in neurolinguistics* (Vol. 1). New York: Academic Press.

Norsk Grunntest for Afasi (Norwegian Basic Aphasia Assessment)

The following is a detailed description of the procedure and content of the test. Procedures and instructions for the patient are given here in English. The test items themselves, however, are given in Norwegian with the English translations in parentheses.

A.1. *Spontaneous Speech*

The patient should, during the course of an interview, be asked three specific questions: "What is your occupation?" "Where do you live?" "What is your favorite TV-program?" He should also be asked to respond to at least two more general questions. Suggestions are "Can you tell me about your family?" "What do you usually do during summer vacation?" The interview should be tape recorded.

A general evaluation of the speech is scored for *communicative efficiency*. The ratings are

0—Normal.

1—The patient expresses his intentions adequately without aid from the examiner. However, his manner of expression is deviant.

2—The patient produces content words and significant clues about his intentions, but some degree of guessing and questioning by the examiner is necessary.

3—The examiner must carry the initiative in order to arrive at a result. The result is questionable, however, because the patient gives too few clues (e.g., responds only yes or no) or because he gives partly contradictory responses.

4—A functional communicative relation is not established.

The next step is to score the presence and the degree of the deviant features that are characteristic of different types of aphasic speech.

Literal Paraphasia

0—Normal.

1—Speech contains distorted words, but these are in most cases interpretable and can be recognized as related to a target word by way of phonemic substitution.

2—A more severe degree of distortion is present, so that interpretability suffers. Grammatical words and filler words are usually not distorted, but content words may be neologistic.

3—Speech consists of *neologistic jargon*. This means varied combinations of syllables and phonemes without recognizable meaning.

Complex Paraphasia

0—Normal.

1—Speech contains semantically related word substitutions or circumlocutory expressions. The target word can be readily guessed.

2—Speech has a topic, but the precise meaning cannot be determined because of bizarre word substitutions and associative leaps in which the target of communication does not seem to be consistently maintained.

3—Uninterpretable jargon is present but mainly with lexically interpretable words.

Visible Effort

0—Normal.

1—Some groping and false starts mark the commencement of an articulatory sequence, but a flow of articulation can then be maintained over several words.

2—Effortful articulation is apparent on almost every word, except on automatized phrases. Effort generalizes to facial musculature and to bodily posture.

3—There is an extensive and general mobilization of physical effort in an attempt to initiate speech. The attempt usually does not get beyond the first syllable or a stereotype.

Hestitation, Pauses

0—Normal.

1—There is some tendency to make unnatural pauses within a sentence.

2—Most sentences acquire an unmelodic and broken character because of frequent pauses.

3—Speech consists of isolated words separated by long pauses.

Stereotypy

0—Not present.

1—An exaggerated use of clichés and empty phrases is apparent.

2—Speech consists exclusively of conventional expressions, for example, swears or polite phrases.

3—Speech consists of an individual stereotype, often a meaningless syllable. The lack of variation distinguishes stereotypy from neologistic jargon.

Articulation (Dysarthria)

0—Not present.

1—Mild. Some slurring of speech occurs, but it does not affect interpretability.

2—Moderate. Phonation or articulation is affected to a degree, making speech difficult to interpret.

3—Severe. A condition of anathria with no or little intelligible speech exists.

Self Correction

0—Normal reaction. The patient corrects slips of the tongue.

1—The patient reacts to paraphasias and usually produces an improved response.

2—The patient reacts to paraphasias but does not produce an improved response.

3—No reaction to severe paraphasia or high degree of stereotypy is present.

The further scoring of spontaneous speech is based on a transcription of the patient's response to the general questions of the interview. Number of words per minute and number of words per utterance are calculated.

A.2. *Auditory Comprehension*

Instruction: "Point to the body part (object) as I name it. Where is the ———?"

Body parts	Objects
1. Tenner (teeth)	1. Klokke (watch)
2. Nese (nose)	2. Ball (ball)
3. Øre (ear)	3. Sikkerhetsnål (safety pin)
4. Panne (forehead)	4. Kopp (cup)
5. Hånd (hand)	5. Nøkler (keys)
6. Hake (chin)	6. Knapp (button)
7. Hals	7. Book (book)
8. Kne (knee)	8. Penn (pen)
9. Lår (thigh)	9. Stein (stone)
10. Albue (elbow)	10. Sokk (sock)
11. Legg (shin)	11. Mynt (coin)

Indirect description (no further instruction given):

Body parts
1. Det du hører med.
(What you hear with.)
2. Det du tramper med.
(What you stamp with.)
3. Det du biter med.
(What you bite with.)

4. Det du lukter med.
(What you smell with.)
5. Det små barn sutter på.
(What babies suck on.)

Objects
1. Det du drikker kaffe av.
(What you drink coffee from.)
2. Det du låser opp med.
(What you use to unlock.)
3. Det barn kaster til hverandre.
(What children throw to one another.)
4. Det du leser i.
(What you read in.)
5. Det du måler tiden med.
(What is used to measure time.)
6. Det du betaler med.
(What you pay with.)

Instruction: "Now I want you to do exactly as I ask."

Body parts
1. Snu deg mot døra.
(Turn toward the door.)
2. Reis deg opp og sett deg ned.
(Stand up and sit down.)
3. Tramp med foten.
(Stamp with your foot.)

4. Strekk ut armen og knytt neven.
(Stretch out your arm and make a fist.)
5. Legg hånden under haka.
(Place your hand under your chin.)

Objects
1. Slå opp boka.
(Open the book.)
2. Trekk opp klokka.
(Wind the watch.)

3. Snu koppen på hodet.
(Turn the cup on its head.)
4. Legg sikkerhetsnåla i koppen.
(Put the safety pin in the cup.)
5. Kast ballen til meg.
(Throw me the ball.)

Body parts	*Objects*
6. Dekk øynene med hånden. (Cover your eyes with your hand.)	6. Løft nøklene og la dem falle. (Lift the keys and drop them.)
7. Klø deg på leggen. (Scratch your shin.)	7. Skyv koppen vekk fra deg. (Push the cup away from you.)
8. Løft benet med hånden. (Lift your leg with your hand.)	8. Legg boka oppå klokka. (Put the book on top of the watch.)
9. Pek nese til meg. (Thumb you nose at me.)	9. Slå opp boka og finn forordet. (Open the book and find the preface.)
10. Pek først på øret, så på kneet og så på albuen. (Touch the ear, knee, and elbow in that order.)	10. Pek først på boka, så på ballen og så på klokka. (Touch the book, ball, and watch in that order.)

Instruction: "Respond yes or no to the following questions. Take time to reflect about them."

Ideas, meaning	*Ideas, relations*
1. Brukes en saks til å klippe med? (Is a pair of scissors used to cut with?)	1. Er en dag kortere enn en uke? (Is a day shorter than a week?)
2. Har kyllinger horn? (Do chickens have antlers?)	2. Er en bjørn større enn en mus? (Is a bear bigger than a mouse?)
3. Er en hest et dyr? (Is a horse an animal?)	3. Er en bestefar eldre enn en gutt? (Is a granddad older than a boy?)

Ideas, meaning	*Ideas, relations*
4. Er Norge et land? (Is Norway a nation?)	4. Er et år lengre enn en måned? (Is 1 year longer than 1 month?)

5. Er kongen en kvinne?
   (Is the king a female?)
6. Har sauer ull?
   (Do sheep have wool?)
7. Er en heks snill?
   (Is a witch good?)
8. Brukes en klut til å
   vispe med?
   (Is a rag used to stir
   with?)
9. Er vann et metall?
   (Is water a metal?)
10. Er en dverg liten?
    (Is a dwarf small?)
11. Brukes en øks til å
    skjære med?
    (Is an ax used to cut
    with?)
12. Er en jolle en båt?
    (Is a dinghy a boat?)
13. Har hunder snute?
    (Do dogs have a snout?)
14. Brukes en kopp til å
    spise av?
    (Is a cup used to eat
    from?)

A.3. Repetition

Instruction: "Please repeat these words (sentences) exactly as I say them."

Words	*Sentences*
1. Mann (man)	1. Fine greier (A fine mess)
2. Bord (table)	2. Sterke saker (Hot stuff)
3. Tre (three or tree)	3. Takk for maten (Thanks for the meal)
4. Femten (fifteen)	4. Ryk og reis (Son of a gun)
5. Tolv (twelve)	5. Sola skinte hele dagen. (The sun shone all day.)
6. Pil (arrow)	6. Du store all verden (What in the world!)
7. Katt (cat)	7. Regnet trommet på taket. (The rain was drum- ming on the roof.)
8. Lys (light)	8. Båten sank i hytt og vær. (The ship sank as the wind blew.)
9. Parafin (kerosene)	9. Han forlangte gjelden betalt. (He demanded payment of the debt.)
10. Syvogtredve (thirty- seven)	10. Aldri annet enn om og men (Never anything except ifs and buts.)
11. Fire tusen (four thousand)	11. Tomme tønner er bedre enn ti på taket. (Empty barrels are bet- ter than 10 on the roof.)
12. Tomater (tomatoes)	
13. Parkere (park)	
14. Skrekk (horror)	
15. Omvende (convert)	
16. Fangst (catch)	

Words	*Sentences*

17. Hundreogtreognitti
    (hund and ninety-three)
18. Skramleorkester
    (junk-instrument
    orchestra)
19. Ansette (hire)
20. Janitsjarkonsert (brass-
    band concert)

Instruction: "These words don't mean anything. Try to repeat them exactly as I say them."

Nonsense syllables

1. ral   5. omlette
2. sob   6. foniter
3. tef   7. balfere
4. gyp   8. maloper

A.4. Naming

Instruction: "Now I point to a body part (object) and you tell me what it is."

Body parts

1. Tenner (teeth)
2. Nese (nose)
3. Øre (ear)

4. Panne (forehead)
5. Hånd (hand)
6. Hake (chin)
7. Hals (throat)
8. Kne (knee)
9. Lår (thigh)
10. Albue (elbow)
11. Legg (shin)

Objects

1. Klokke (watch)
2. Ball (ball)
3. Sikkerhetsnål (safety
   pin)
4. Kopp (cup)
5. Nøkler (keys)
6. Knapp (button)
7. Bok (book)
8. Penn (pen)
9. Stein (stone)
10. Sokk (sock)
11. Mynt (coin)

Instruction: "Describe what I am doing."

Body parts	Objects
1. Reise seg. (Stand up.)	1. Lese. (Read.)
2. Knytte neven. (Make a fist.)	2. Kaste ball. (Throw a ball.)
3. Trampe. (Stamp your foot.)	3. Trekke klokka. (Wind the watch.)
4. Dekke øynene med hånden. (Cover your eyes with the hand.)	4. Snu koppen. (Turn the cup around.)
5. Klø seg på leggen. (Scratch your shin.)	5. Legge sikkerhetsnåla i koppen. (Put the safety pin in the cup.)

Instruction: "Respond to the following questions. A brief answer is sufficient."

1. Hvilken farge har snø? (What is the color of snow?)
2. Hvem bor på slottet? (Who lives in the royal castle?)
3. Hvor mange dager er det i en uke? (How many days are there in a week?)
4. Hva heter den første måned i året? (Which is the first month of the year?)
5. Hva brukes såpe til? (What is soap used for?)
6. Hva brukes en saks til? (What is a pair of scissors used for?)
7. Hva brukes en blyant til? (What is a pencil used for?)
8. Hvor mange kilometer er det i en mil? (How many yards make a mile?)
9. Hva bruker man å hogge ved med? (What do you use to chop wood?)
10. Hvilken smak har sitroner? (How do lemons taste?)

A.5. Reading

Instruction: "Read what is printed on this card" (reading aloud); "Point to the letter or word that I say" (comprension); "Read the card and find the object" (comprehension); "Read the card and do what it says" (comprehension).

Reading aloud	Comprehension
1. B	1. B
2. A	2. A
3. R	3. R
4. O	4. O
5. Y	5. Y
6. K	6. K

1. Ball (ball)	1. Ball
2. Bok (book)	2. Bok
3. Kopp (cup)	3. Kopp
4. Klokke (watch)	4. Klokke
5. Nøkler (keys)	5. Nøkler
6. Sikkerhetsnål (safety pin)	6. Sikkerhetsnål
7. Tomater (tomatoes)	
8. Jugoslavia (Yugoslavia)	
9. Sentimental (sentimental)	
10. Aktivitet (activity)	

1. Trekk opp klokka. (Wind the watch.)	1. Trekk opp klokka.
2. Lukk øynene. (Close your eyes.)	2. Lukk øynene.
3. Kast ballen til meg. (Throw me the ball.)	3. Kast ballen til meg.
4. Klø deg på leggen. (Scratch your shin.)	4. Klø deg på leggen.
5. Rør først ved ballen, så ved boka og så ved klokka. (Touch the ball, the book, and the watch in that order.)	5. Rør først ved ballen, så ved boka og så ved klokka.

A.6. Syntax

Instruction: "The words on these three cards make up a sentence. Try to arrange them in the correct order. Do not try to make questions."

1. Festen varte/ til langt/ på natt/
   (The party lasted long into the night.)
2. Flyet alle/ ventet på/ kom ikke/
   (The plane everyone waited for did not come.)
3. Brevet jenta/ skrev forsvant/ i posten/
   (The letter the girl wrote got lost in the mail.)
4. Store sultne/ løver brølte/ etter mat
   (Big, hungry lions roared for food.)
5. Buksa damen/sydde til/sønnen passet/
   (The pants the woman sewed for her son fitted.)
6. Før dagen/var slutt/kom snøen
   (Before the day was over the snow came.)

A.7. Writing

Instruction: "Sign your name;" "Copy these words."

1. Kopp (cup)
2. Sikkerhetsnål (safety pin)

"Write the words I say."

1. Ball (ball)
2. Klokke (watch)

"What is this? Write down the name."

1. Nøkler (keys)
2. Øre (ear)

"Write the following sentence."

1. Klø deg på leggen. (Scratch your shin.)
2. Båten sank i hytt og vær. (The ship sank as the wind blew.)

Index